Science, Technology, and Virtues

Science, Technology, and Virtues

Contemporary Perspectives

Edited by

EMANUELE RATTI
AND
THOMAS A. STAPLEFORD

OXFORD
UNIVERSITY PRESS

OXFORD
UNIVERSITY PRESS

Oxford University Press is a department of the University of Oxford. It furthers the University's objective of excellence in research, scholarship, and education by publishing worldwide. Oxford is a registered trade mark of Oxford University Press in the UK and certain other countries.

Published in the United States of America by Oxford University Press
198 Madison Avenue, New York, NY 10016, United States of America.

© Oxford University Press 2021

Library of Congress Cataloging-in-Publication Data
Names: Ratti, Emanuele, editor. | Stapleford, Thomas A., editor.
Title: Science, technology, and virtues : contemporary perspectives /
edited by Emanuele Ratti and Thomas A. Stapleford.
Description: New York : Oxford University Press, [2021] |
Includes bibliographical references and index.
Identifiers: LCCN 2021015587 (print) | LCCN 2021015588 (ebook) |
ISBN 9780190081713 (hb) | ISBN 9780190081744 | ISBN 9780190081720 |
ISBN 9780190081737 (epub)
Subjects: LCSH: Research—Moral and ethical aspects. | Science—Philosophy. |
Technology—Philosophy.
Classification: LCC Q180.55.M67 S34 2021 (print) | LCC Q180.55.M67 (ebook) |
DDC 174/.95—dc23
LC record available at https://lccn.loc.gov/2021015587
LC ebook record available at https://lccn.loc.gov/2021015588

DOI: 10.1093/oso/9780190081713.001.0001

1 3 5 7 9 8 6 4 2

Printed by Integrated Books International, United States of America

Contents

I. VIRTUES IN SCIENCE

II. VIRTUES AND TECHNOLOGY

III. VIRTUES AND EPISTEMOLOGY

IV. VIRTUES AND RESEARCH ETHICS

Acknowledgments

This volume emerged from a workshop held at the University of Notre Dame on April 5–7, 2018, that was supported both by the Reilly Center for Science, Technology, and Values and by a project from the Templeton Religion Trust, "Developing Virtues in the Practice of Science." The latter played a formative role in helping both editors think more deeply about virtues, how they might function in scientific research, and how they relate to science and technology studies more broadly. We are very much in debt to the core project team for their intellectual help, administrative support, and friendship over the three years of the project, including Celia Deane-Drummond, Darcia Narvaez, Becky Artinian-Kaiser, Dori Beeler, Louise Bezuidenhout, Emily Dumler-Winckler, Tim Reilly, Katie Rutledge, Fionagh Thomson, and Nate Warne. We are also grateful for the advice, feedback, and comments from all participants in the April 2018 workshop and for the comments of the anonymous reviewers for Oxford University Press. Finally, John Nagy provided invaluable editorial assistance on multiple chapters from the volume and made the book much stronger and eloquent as a result.

Editors

Contributors

Jon Alan Schmidt, PE, SECB, is a senior associate structural engineer in the Aviation & Federal Group at Burns & McDonnell in Kansas City, Missouri. He is a frequent author and speaker on philosophical aspects of engineering practice, was the founding chair of the Engineering Philosophy Committee within the Structural Engineering Institute of the American Society of Civil Engineers, and served on the international steering committee for the Forum on Philosophy, Engineering and Technology.

Dori Beeler is a medical anthropologist interested in health disparities and well-being. She is currently a postdoctoral fellow at Johns Hopkins Bloomberg School of Public Health working in pediatric cancer prevention and control.

Richard Bellon is the author of *A Sincere and Teachable Heart: Self-Denying Virtue in British Intellectual Life, 1736–1859* (Brill, 2015). He is Associate Professor with a joint appointment in Lyman Briggs College and the Department of History at Michigan State University, where his research focuses on nineteenth-century British science.

Louise Bezuidenhout is senior researcher in the Department of Information Systems, University of Cape Town. She is a social scientist specialising in Critical Data Studies. Her work examines the evolving Open Data/Open Science landscape and the evolution of data sharing infrastructures, practices and communities. She has also worked for many years in the field of Responsible Research, and uses virtue ethics to rethink how broad discussions of responsibility can be linked to daily research practices.

Mark Bourgeois is Assistant Teaching Professor in Notre Dame's Technology Ethics Center. A philosopher and former engineer, he has interests in technology ethics as well as the ethics education of scientists and engineers. He co-directed the Social Responsibilities of Researchers project in the Reilly Center for Science, Technology, and Values of the University of Notre Dame and currently directs the Responsible Innovation Fellowship program.

Jiin-Yu Chen is Lead of Research Compliance at Baylor College of Medicine. She was formerly Director of Research Integrity and Postdoctoral Affairs at Boston College, where she oversaw the areas of responsible conduct of research, conflict of interest, and research misconduct. Her research interests include developing best practices for responsible conduct of research education programs.

Daniel J. Hicks is a philosopher of science and Assistant Professor in the Department of Cognitive and Information Sciences at the University of California, Merced. Their research interests focus on the role of ethical and political values in science and public scientific controversies.

Mike W. Martin, Emeritus Professor of Philosophy at Chapman University, has published almost one hundred articles and sixteen books, primarily on virtue ethics in science and engineering. His award-winning textbook *Ethics in Engineering* has gone through four editions and been translated into five other languages.

Jon D. Miller is a political scientist in the Institute for Social Research at the University of Michigan, where he serves as Director of the International Center for the Advancement of Scientific Literacy. He has studied public understanding of science and technology for nearly five decades, and his measure of civic scientific literacy has been used in more than forty countries.

Robert T. Pennock is University Distinguished Professor of History, Philosophy and Sociology of Science at Michigan State University, where he is a principal investigator of the Scientific Virtues Project and co-principal investigator of the BEACON Center for the Study of Evolution in Action. A Fellow of the American Association for the Advancement of Science and a National Associate of the National Academies of Science, he is the author of *An Instinct for Truth: Curiosity and the Moral Character of Science* (MIT Press, 2019).

Laura Ruetsche is Louis Loeb Collegiate Professor of Philosophy at the University of Michigan, where she studies philosophy of physics, general philosophy of science, and feminist philosophy. Ruetsche has held fellowships from the ACLS and the Center for Advanced Study of the Behavioral Sciences and was the 2013 winner of the Lakatos Award in the philosophy of science.

Jutta Schickore is Ruth N. Halls Professor and Chair of the Department of History and Philosophy of Science and Medicine at Indiana University, Bloomington. The author of two books and co-editor of three more, Schickore works on the philosophical and scientific debates about scientific methods in the past and present. She has held fellowships at the Wellcome Institute, the Dibner Institute, the National Humanities Center, the Princeton Institute for Advanced Study, and the Max Planck Institute for the History of Science.

John P. Sullins is Professor of Philosophy at Sonoma State University. He has published extensively on the ethics of emerging technologies involving artificial intelligence, robotics, and the ethics of information technologies and was the 2011 winner of the Herbert A. Simon Award for Outstanding Research in Computing and Philosophy. He is the co-author of *Great Philosophical Objections to Artificial Intelligence: The History and Legacy of the AI Wars*.

Dana Tulodziecki is Associate Professor of Philosophy at Purdue University, where she works on the philosophy of science and the role of epistemic virtues in scientific practice. Her papers have appeared in *Philosophy of Science*, the *British Journal for Philosophy of Science*, *Studies in History and Philosophy of Science*, and *Synthese*, among others.

Shannon Vallor is the Baillie Gifford Chair in the Ethics of Data and Artificial Intelligence at the University of Edinburgh, where she is appointed in Philosophy and directs the Centre for Technomoral Futures at the Edinburgh Futures Institute. The author of *Technology and the Virtues* (Oxford University Press, 2016), Vallor is editor of the forthcoming *Oxford Handbook of Philosophy of Technology* and the 2015 recipient of the World Technology Award in Ethics.

Using Virtue to Think About Science and Technology

Thomas A. Stapleford and Emanuele Ratti

In the twenty-first century, talk of virtue can conjure images of Victorian patriarchy or dusty children's books filled with earnest morality tales about honesty or courage. Against that background, it may seem implausible to propose virtue as a valuable concept for thinking about science and technology, those quintessentially late modern phenomena. And yet over the last two decades, scholars across multiple disciplines have done just that, arguing that the lens of virtue can illuminate and guide both the pursuit of science and the design and use of technology (e.g., Daston and Galison 2007; Shapin 2009; Fairweather 2014; Vallor 2016; Dongen and Paul 2017; Pennock 2019; Scherz 2019). This volume gathers a variety of such perspectives to show how virtue concepts can enrich our understanding of scientific research, guide assessments of the design and use of new technologies, and, we hope, inform our decisions about how we form scientists, engineers, consumers, and citizens writ large. At the heart of the book is a conviction that science and technology ought to be oriented toward the good life, and that virtue theory can help us pursue that orientation.

Science, Technology, and Virtue: A Brief History

Thinking about virtue in relationship to science would not have seemed incongruous for much of Western history. From ancient Greece through the Middle Ages, philosophers routinely promoted the study of the natural world as a means for cultivating intellectual and moral virtues. Indeed, the Latin word *scientia* referred primarily not to a body of knowledge or to an activity but to the ability to reason properly from first principles—that is, to an intellectual virtue (Harrison 2015, esp. ch. 2). Such claims continued into the early modern period (e.g., Jones 2006; Corneanu 2011), even as *scientia* began to take on its modern connotations and to be justified on other grounds, such as practical utility (Harrison 2015, esp. chs. 4 and 5). Still, although the development of virtue became less a rationale for studying the natural world than a commonly perceived prerequisite for

Thomas A. Stapleford and Emanuele Ratti, *Using Virtue to Think About Science and Technology* In: *Science, Technology, and Virtues*. Edited by: Emanuele Ratti and Thomas A. Stapleford, Oxford University Press. © Oxford University Press 2021. DOI: 10.1093/oso/9780190081713.003.0001 successful scientific work, the links between virtue and science remained strong through much of the nineteenth century (e.g., Bellon 2014).

The ties were eroding, however, and by the middle of the twentieth century, explicit talk about virtue had largely disappeared from discussions of science. Much of this decline surely owed to the broader decline of virtue as a central category in moral philosophy by the early twentieth century. At least in academic circles, discussions of character or virtue in moral philosophy were supplanted by talk of rules, principles, duties, and consequences—a shift whose exact contours and causes remain contentious (Frede 2013). But some of the eroding bonds between science and virtue can also be traced to changes in science itself. Though much work remains to be done on this topic, two common themes emerge from the best broad analyses of these shifts, those by Shapin (2009) and Harrison (2015).

First, the social organization of science became more formal, elaborate, and tightly organized through hierarchical structures with a substantial division of labor (laboratories, universities, corporate research divisions, government agencies, etc.). With behavior governed by bureaucratic rules and social norms, the personal characteristics of individual scientists, and certainly their moral qualities, could readily be seen as irrelevant to the successful production of knowledge. Far from viewing science as cultivating or requiring particular virtuous traits, many twentieth-century commentators were keen to emphasize what Steven Shapin has called the "moral equivalence of the scientist"—scientists were just like other people, neither better nor worse (2009, 47–91).

Second, the early modern period marked the beginning of a divorce between science and metaphysics, a split that was never entirely complete but which became increasingly accepted by scientists, especially by the early twentieth century. Instead of providing access to a singular reality (one that early modern philosophers saw as reflecting the mind of God), scientific knowledge became a tool for manipulating the material world. With science no longer linked to the development of moral character and having been stripped of its theological-metaphysical ambitions, talk of virtue no longer seemed relevant to many practitioners. What, after all, was especially virtuous about the pursuit of power? Indeed, one might expect precisely the opposite. As David Hull concluded from his study of biologists: "The least productive scientists tend to behave the most admirably, while those who make the greatest contributions just as frequently behave the most deplorably" (Hull 1988, 32).

If aspects of these two trends—the disappearance of the individual and the dominance of the pursuit of power—can be located much earlier (notably in Francis Bacon, for example), they nonetheless culminated during World War II as scientists were gathered into massive collective projects for military research.

The Manhattan Project alone put to rest any simplistic alignment of science with virtue, and the embrace of scientific research by Cold War states and corporate capitalism in the postwar era only reinforced the seemingly amoral character of science and scientists. Science could be used for noble purposes or base ones; there was nothing inherent in science or scientists themselves to pull primarily in one direction or the other.

It is perhaps no surprise that the era that saw the demise of explicit ties between science and virtue also exemplified the dominance of what is often called *technoscience*—namely, the full integration of science and technology. As an analytic term, technoscience can direct our attention to science's inescapable reliance on material artifacts for studying and intervening in the natural world, a feature that was as much a part of ancient natural philosophy as it is of contemporary science (Hottois 2006). But as a descriptive term, technoscience captures a potentially much tighter interconnection between scientific and technological development, as scientific innovation increasingly both demands and facilitates technological innovation and thus the boundaries between science and engineering become deeply blurred. It is technoscience in this latter respect that has come to characterize the study of the natural world since the early twentieth century and to parallel the breakdown of a nexus between science and virtue.

Unlike science, the construction of technology has always had an ambivalent relationship to virtue. For example, although Aristotle famously used *techne*—the intellectual disposition necessary for successful *poiesis*, or making—as an analogy for moral virtue, he also took great pains to distinguish them and to insist that procuring the one had no connection to developing the other (Russell 2014). Thus whereas both Plato and Aristotle understood the study of the natural world as a pathway to cultivating moral virtue and leading the intellect to the divine, they (and many of their ancient and medieval successors) saw the pursuit of *techne* as a base occupation. Of course, medieval and Renaissance authors often took a more positive view of artisans and did see personal qualities as tied to successful technological innovation, usually under terms such as *ingenuity, cunning,* or *wit* (Marr et al. 2019). But, just as with science, the collective and highly structured approach to large engineering projects that spread during the twentieth century destabilized the place of the individual genius. The contrast was well illustrated by the failure of Thomas Edison's military research projects during World War I in comparison to the success of newer industrial labs grounded in academic engineering and scientific research (Kevles 1977, 8–9, 116, 136–38). Like the pursuit of science, the development of technology was seemingly neither an inherently moral activity nor dependent on the personal qualities of an individual beyond the possession of technical skills and knowledge.

The Return of Virtue?

Despite these trends, a new and growing group of scholars have found virtue to be a valuable concept for thinking about twentieth-century science and technology. This resurgence of interest in virtue no doubt has partial roots in the revival of virtue ethics—often traced to G. E. M. Anscombe's essay "Modern Moral Philosophy" (1958)—and in the growth of virtue epistemology. But its heart lies in the "practice turn" of science and technology studies (STS) from the 1970s through the 1990s (e.g., Pickering 1992; Schatzki, Knorr-Cetina, and Savigny 2001; Soler 2014). Briefly put, the practice turn entailed moving the study of science and technology away from a focus on ideas and artifacts, respectively, and toward the activities that constructed, sustained, and employed them. As attention shifted from things to actions, science and technology studies evolved in ways that resonated with the revitalization of virtue ethics and virtue epistemology. Suddenly, to speak of virtue in relation to contemporary science and technology no longer seemed antiquarian but was perceived as fresh and illuminating.

First, as scholars looked closely at scientists' actions, they began to reveal how being a scientist demanded more than following sets of explicit rules or techniques (Polanyi 1964); it entailed cultivating judgment and forms of self-discipline (Daston and Galison 2007), building habits of mind and behavior (Warwick 1998), and even developing proper emotional responses (White 2009). In light of Wittgenstein's (1963) arguments about the limits of formal rules and Michel Foucault's studies of institutional and social practices as ways of crafting the self (e.g., 1995), the familiar Weberian view of modern social organization effacing the personal qualities of individuals was turned on its head. Far from institutional practices and social norms supplanting the character and traits of scientists, they were recast precisely as tools for fostering character and personal traits. Only properly formed dispositions would allow scientists to respond appropriately to contingent situations, scenarios whose complexities could never be captured by explicit rules (consider Warwick and Kaiser 2005).

Initially, scholars rarely described these dispositions as "virtues"—the language had been lost, and the vicious behavior of many putatively successful scientists still kept the term at bay. But closer studies revealed inescapable ethical dimensions to many of the personal qualities promoted within science (e.g., Herzig 2005). As Steven Shapin (2009) argued at length, beneath the superficial twentieth-century narrative of amoral and impersonal technoscience lay a vibrant counternarrative wherein success was deeply dependent on the moral qualities and personal characteristics of its practitioners. For every account of science or engineering as staid fields populated by impersonal conformists with pocket calculators, there was a counterstory of a bongo-playing Richard

Feynman or a brash Silicon Valley entrepreneur, and both grant decisions and capital investments relied heavily on judgments about the character and qualities of key individuals.

Meanwhile, the revival of virtue ethics and virtue epistemology created a richer conception of virtue that rendered it more amenable to the messy realities depicted in science and technology studies. Scholars were reminded that *virtue* originally denoted any disposition deemed to be excellent or admirable, and it could therefore encompass intellectual or sensorimotor qualities that might be (in principle) separable from moral virtue. Moral virtues themselves could be seen as ideals, not descriptions of every practitioner, and virtue ethicists insisted that the effects of their absence could be assessed only through a proper ethical lens; vice was perfectly compatible with success defined in narrow terms of fame or power. In this context, some STS scholars began to talk explicitly about virtue in relation to aspects of twentieth-century science such as objectivity (Daston and Galison 2007), creativity (Martin 2007), trust and collaboration (Shapin 2009; Frost-Arnold 2013), epistemic judgment (Stump 2007; Fairweather 2014), and interlocking epistemic and ethical debates (Wang 2017; Bont 2017).

If contemporary science is indeed technoscience, then it should be no surprise that the practice turn has similarly drawn attention to virtue in accounts of technology more generally. Scholars have talked about the relevance of virtues to design and to the training of engineers (Harris 2008; Martin and Schinzinger 2010, 60–73; Stovall 2011); how a virtue framework might affect our use and assessment of technology (Elder 2014; Frost-Arnold 2014; Vallor 2016); and even how virtues could or should be embodied within robots and other artificial agents (see essays by Howard, Vallor, and Sullins in Floridi and Taddeo 2014).

Yet despite the breadth of this recent scholarship on virtue, much of it has occurred in isolated pockets or subfields, and thus it has been easy to overlook the resurgence of interest in virtue across science and technology studies. In this volume, we have worked against that trend, aiming to capture both the vibrancy and the diversity of virtue-based literature. Toward that end, we have solicited both new research and revised versions of classic essays that (collectively) examine virtue in relation to both science and technology. Our contributors include senior scholars, early-career academics, and practitioners in engineering and research ethics training who come from a broad range of disciplinary backgrounds: philosophy, history, sociology, anthropology, political science, and engineering.

Given this diversity, it is no surprise that our contributors draw on different conceptions of virtue and connect it to science in different ways. For example, most of our authors follow traditional views in which traits count as virtues only if they are constitutive of being an excellent human being writ large. But others, such as Pennock and Miller, adopt a narrower definition in which virtues are specific to success in particular activities. Or again, as in the revival of virtue ethics more broadly, several of our authors (such as Schmidt, Chen, and Stapleford and Hicks) draw heavily on a loosely Aristotelian framework. But others, notably Vallor and Martin, pull from a wider range of cultural and intellectual traditions. Though most of our authors treat virtues solely as traits of human beings, others push us to consider virtuous technologies (Sullins) or virtuous theories (Tulodziecki). Finally, the different methodologies our contributors employ necessarily shape what they can, and cannot, say about virtue. For example, since it is impossible to measure traditional conceptions of virtue directly, efforts to study virtue empirically (such as in essays by Pennock and Miller, Bourgeois, and Bezuidenhout and Beeler) must come at the topic indirectly.

Despite this diversity, our contributors are united in their conviction that virtue can be a valuable concept for understanding modern science and technology. In this respect, the essays in this volume reflect the growing scholarship around virtue—an excitement about the insights that it can afford even as scholars take their work down diverging paths. Though continued progress may eventually demand consolidation and the construction of sharper boundaries, that time has not yet come. Therefore, rather than trying to force our contributors into an artificially imposed unanimity, we have opted to showcase the pluralism of research on virtue in science and technology with the hope that others will find aspects of virtue-based scholarship that resonate with their own work and which they can develop in their own ways.

Our volume is divided into four major sections. The first, "Virtues in Science," offers a general examination of how both moral and intellectual virtues have, and continue to be, relevant to scientific practice. Richard Bellon's essay, "Sacrifice in Service to Truth: The Epistemic Virtues of Victorian British Science," opens the volume by taking us back to the nineteenth century, a time when the language of virtues was still an explicit and common part of scientific discourse. Bellon shows how epistemic virtues from Victorian Britain—virtues believed to aid in constructing reliable knowledge—relied heavily on moral traits associated with self-denial, such as patience, perseverance, and humility. Bellon argues that these moralized epistemic virtues had their roots in Christian theological ethics. By the end of the nineteenth century, however, they had been rhetorically disconnected from Christianity, serving instead as a secularized morality grounded in what was perceived as good scientific practice. Bellon tracks this transition through a case study of how the virtue language common to the University of Cambridge became both "shield and sword" in Victorian debates over evolution, eventually being adopted even by agnostics such as Thomas Huxley and Charles Darwin himself.

In "Seeing Science as a Communal Practice: MacIntyre, Virtue Ethics, and the Study of Science," Tom Stapleford and Dan Hicks explore how the moral philosopher Alasdair MacIntyre's concept of a "practice" can provide a valuable framework for understanding contemporary science. Building on previous work (Hicks and Stapleford 2016), they show how MacIntyre's development of Aristotelian virtue ethics can help us understand much of the moral rhetoric around science, such as debates about the "purity" of science or about whether science is a vocation (a calling with moral overtones) or merely a job (a means of making money). Rather than treating these as objective descriptions of scientific research, Stapleford and Hicks argue that they are competing normative narratives about the goals and character of scientific practice. To see science as a communal practice is to place scientific practice within a moral narrative wherein the rightly ordered pursuit of knowledge develops scientists' virtues while also contributing to the good of their communities. Stapleford and Hicks insist that it is possible, indeed vital, to hold on to that perspective even while recognizing the complex motivations of scientists and the frequent disconnect between moral virtues and what may be perceived as "successful" scientific research.

For our third essay, "Studying Scientific Virtues: Bridging Philosophy and Social Science," the philosopher Robert T. Pennock and the political scientist Jon Miller reflect on the challenges and prospects of using social-scientific tools to inform the philosophical analysis of science—in this case, the relevance of virtue to science. Pennock and Miller remind us that even as scholars have been using virtue concepts to think about science and social scientists have been conducting empirical studies of moral development and ethical reasoning, understanding the role of virtue in science requires collaboratively bridging the wall that usually separates these disciplines. Turning to their own collaboration, Pennock and Miller describe their ongoing efforts to examine what character traits contemporary scientists value. As they argue, surveys and statistical analyses can inform philosophical claims by revealing the fine-grained differences in what Pennock calls "vocational virtues" across multiple professions (including science).

Our second section, "Virtues and Technology," provides a parallel introduction to how virtues can help us think effectively about the practices of using and designing technology. Shannon Vallor's opening essay sets the stage by providing an overview of themes she developed in her monograph *Technology and the Virtues: A Philosophical Guide to a Future Worth Wanting* (2016). Vallor argues that emerging technologies of both the recent past and the near future have highlighted the salience of "technomoral choices"—ethical choices about how we engage with and develop the elaborate technological systems that increasingly mediate our lives. Though humans have always faced such choices, we are now in an era where the effects of technological change can be more rapid, more massive in scale and impact, and more difficult to predict than ever. In Vallor's view, this dynamism and uncertainty have undermined the effectiveness of traditional approaches to ethics, such as principle- or rule-based ethics and consequentialism. Though Vallor sees no easy solution to this dilemma, she proposes cultivating what she calls "technomoral virtues" (drawn from virtue ethics traditions in Aristotelian, Buddhist, and Confucian ethics) as a strategy for making ethical decisions in an unpredictable and rapidly changing technological environment.

In "Mindful Technology," Mike W. Martin takes up this challenge with a reflection on one potential technomoral virtue, mindfulness. For Martin, mindfulness entails "paying attention to what matters in light of relevant values," but as he describes it (drawing on both Western and Buddhist traditions), mindfulness can manifest itself in a variety of forms. A similar diversity emerges in how scholars and social commentators have promoted mindfulness as an ethical value for media use, creativity in design, responsible engineering, and good citizenship. As Martin emphasizes, clarity about what we mean by "mindfulness" is critical to any ethical evaluation, and he argues that mindfulness can become a virtue only if the values to which we are attentive are linked to broader conceptions of individual and communal good.

Jon Alan Schmidt examines the relevance of virtues to engineering in his essay, "Virtuous Engineers: Ethical Dimensions of Technical Decisions." Returning to ancient Greek categories of *techne* (knowledge of how to make something) and *phronesis* (practical judgment for living rightly), Schmidt acknowledges that good engineering is often assumed to rely primarily on *techne*. But he argues that this view is mistaken: instead of applying fixed rules to reach clearly defined ends, engineers must routinely rely on practical judgment to discern what constitutes a good end—a good design—in each particular context. In Schmidt's depiction, engineering matches Alasdair MacIntyre's concept of a "practice" (the topic of Stapleford and Hicks's essay) and therefore necessarily demands the cultivation of moral and intellectual virtues. Virtuous engineers are not merely technicians who create whatever they are told to construct; instead, they pursue the well-being of all humans through the internal goods of engineering design (safety, sustainability, and efficiency).

Whereas Schmidt considers the necessity of virtues for designing good technology, John Sullins considers how and why we might wish to instill virtues in our technology. In "Artificial Phronesis: What It Is and What It Is Not," Sullins delineates what would be entailed in ascribing the central Aristotelian virtue of *phronesis* to artificial agents and why we ought to pursue such a goal, despite its difficulty. Sullins distinguishes ethical impact agents (EIAs, whose actions have ethically significant consequences but who engage in no ethical reasoning) and artificial ethical agents (AEAs, who follow an ethical calculus explicitly programmed into them) from what he calls artificial moral agents (AMAs). In Sullins's classification, EIAs and AEAs have no direct moral culpability; they can be simple or complex learning machines, but they only follow their explicit preprogrammed goals. By contrast, an AMA "is either a conscious moral agent or functionally equivalent to one" because it operates via practical reasoning in a manner similar to human beings. Accordingly, it demands a parallel form of *phronesis*—namely, artificial *phronesis* (AP). Though AP may be a lofty and even unreachable goal, Sullins argues that our increasing dependence on technology to mediate our lives makes the pursuit of AP essential.

The chapters of Part III, "Virtues and Epistemology," discuss appeals to virtue within epistemological discussions in the philosophy of science. Emanuele Ratti provides a broad overview of the virtue discourse in this discipline. First, the main use of the word *virtue* in philosophy of science refers not to attributes or traits of agents but rather to the products of science. For instance, there are virtues of theories, models, explanations, and so on. This conception of virtues lies at the core of long-standing debates in the discipline, most notably theory choice and underdetermination. Next, there is a small but growing literature in philosophy of science that "contaminates" traditional concerns of the discipline with insights from virtue epistemology. Ratti describes these recent developments and introduces the field of virtue epistemology as well. He ends the chapter by noticing how the widely debated concepts of epistemic injustice can be profitably connected to topics in feminist philosophy of science and science and values.

In "Virtue and Contingent History: Engineering Science," Laura Ruetsche criticizes a certain way of thinking about (scientific) rationality that she calls "traditionalist epistemology." Exemplified by certain forms of Bayesian subjectivism, traditionalist epistemologies conceive of epistemic warrant as characterized by adherence to a constricted calculus of rationality. Ruetsche thinks that the Aristotelian notion of virtue as second nature can help us characterize what is missing in this traditionalist picture of rationality. Drawing on philosophers such as Hacking, Cartwright, and Galison, who emphasize the epistemic importance of the context of discovery, Ruetsche shows how discovery often demands the kind of second nature that cannot be encompassed within traditionalist epistemologies. Ruetsche explicitly connects this message to feminist philosophy of science and closes the chapter by discussing how a scientific community should be organized so that it might maximize the effective use of epistemically relevant second natures by enlarging Longino's ideal of "cognitive democracy."

In the chapter "Is 'Failing Well' a Sign of Scientific Virtue?," Jutta Schickore proposes a sophisticated analysis of the concept of failure in science. She analyzes the sense in which a failure could be beneficial, given the recent attention to the notion of "desirable failure." In educational contexts, failures can play a pivotal role when they are turned into "teaching moments," and science education is no exception. Thinking about scientific practice in terms of success and failure has surplus value because it prompts us to acknowledge that scientific practice is done by humans and takes place in a social-institutional context. Nevertheless, shifting the emphasis from valuing success to valuing failure is not illuminating. In the second part of the chapter, Schickore illustrates this further by describing her own empirical research project, where scientists are interviewed on their perception of what constitutes a failure in science and whether it can be beneficial. Most interviewees admit that failure is perceived as a personal defeat, and in general as a stigma. Interviewees also comment on negative results, rather than just on failures in a scientific project as a whole, and it turns out that negative results are as stigmatized as failure. Schickore also notes a slippage from cognitive or objective to moral failure "in the context of comments about the failure to replicate something." She concludes the analysis of interviews by noticing that interviewees conceptualize scientific practice almost exclusively in terms of "failure" and "success," which is problematic because it appears to be an obstacle to effective peer criticism and open discussion of methodological issues in established scientific fields.

Finally, Dana Tulodziecki discusses the relation between virtues and science from a different point of view. She looks at the problem of virtues in science in the context of underdetermination, scientific realism, and theory choice. Tulodziecki emphasizes the virtues of scientific theories and how we might be able to choose among competing scientific theories on this basis. Proposing to deal with the theoretical virtues empirically, she investigates the epistemic import of virtues through examining how they work in the history of science. She describes in detail the case of puerperal fever in mid-1800s Britain, with a focus on how virtues were used by participants in the debate about its transmissibility. Tulodziecki concludes that her case suggests that theoretical virtues are sometimes epistemic, even though not necessarily truth-conducive. Most importantly, her concern is not with whether specific virtues are or are not epistemic but with showing that it makes sense to settle this question empirically.

In Part IV, "Virtues and Research Ethics," chapters discuss how a virtue perspective can be useful in shaping a new way of conceptualizing and teaching research ethics. Jiin-Yu Chen in "Integrating Virtue Ethics into Responsible-Conduct-of-Research Programs: Challenges and Opportunities" makes a thorough analysis of responsible-conduct-of-research (RCR) programs in the United States and how they have been and could be shaped by virtue ethics perspectives broadly conceived. By drawing from concrete examples of different RCR programs, she first describes in detail the obstacles that RCR programs usually face. Her analysis suggests that there is a tension in these programs—while RCR programs emerge from the need for compliance and are increasingly shaped by this, their goals in principle are aspirational, in the sense of aspiring to help researchers to cultivate ethical judgment capabilities. Next, Chen introduces virtue ethics into the analysis. She describes some programs that have prioritized virtue ethics over other approaches (such as work by Pennock), but she urges that we should incorporate a virtue ethics component into preexisting RCR programs and reflects on the critical role mentoring relationships play in cultivating virtues.

In "Virtue Ethics and the Social Responsibilities of Researchers," Mark Bourgeois explores a virtue ethics approach to teaching social responsibility in research integrity/RCR programs. Bourgeois describes in detail a project developed with the philosopher Don Howard and funded by the US National Science Foundation. After pointing out the different ways in which innovations and research outputs can affect society, Bourgeois emphasizes that conventional ethics training lacks a proper characterization of potential social responsibilities of research, as well as a more holistic view of research that includes the ethical role of researchers and innovators. Given these limitations, Bourgeois highlights the value of a virtue ethics approach, where cultivation of moral development becomes fundamental for tackling very ambiguous and messy issues of social responsibility.

Finally, in "Dynamic Boundaries: Using Boundary Work to Rethink Scientific Virtues," Louise Bezuidenhout and Dori Beeler analyze how conceptions of what constitutes science and the scientist can have important consequences for research ethics. In particular, they emphasize that the image of the scientist as being "other" from society persists even today. Sometimes science and scientists are identified by appealing to essentialist criteria, thereby separating them from society. These representations are ethical, or have ethical consequences, because they are used to shape the relation between science and society and what each owes to the other. The authors describe literature in STS that challenges these ideas, in particular by showing that "instead of static space with certain essential practices/traits, science is defined by what ever-fluctuating boundaries enclose or exclude." In consequence, the ethics of science cannot be encapsulated into fixed rules, because nothing is fixed in science.

Taken together, these essays do not offer a simple panacea for improving modern science and technology, or even a unified view of virtue. But they do illustrate how virtue concepts can provide a productive and illuminating perspective on two enterprises at the core of modernity, pressing us to see how intellectual and moral character—as embodied dispositions for action—continue to be central for pursuing the good life, even in an age of high technology and science.

References

Anscombe, G. E. M. 1958. "Modern Moral Philosophy." *Philosophy* 33: 1–19.
Bellon, Richard. 2014. *A Sincere and Teachable Heart: Self-Denying Virtue in British Intellectual Life, 1736–1859.* Leiden: Brill.

Bont, Raf. 2017. "The Adventurer and the Documentalist: Science and Virtue in Interwar Nature Protection." In *Epistemic Virtues in the Sciences and the Humanities*, edited by Jeroen van Dongen and Herman Paul, 129–147. Cham, Switzerland: Springer.

Corneanu, Sorana. 2011. *Regimens of the Mind: Boyle, Locke, and the Early Modern Cultura Animi Tradition.* Chicago: University of Chicago Press.

Daston, Lorraine, and Peter Galison. 2007. *Objectivity.* Cambridge, MA: Zone Books.

Dongen, Jeroen van, and Herman Paul, eds. 2017. *Epistemic Virtues in the Sciences and the Humanities.* Cham, Switzerland: Springer.

Elder, Alexis. 2014. "Excellent Online Friendships: An Aristotelian Defense of Social Media." *Ethics and Information Technology* 16, no. 4: 287–97.

Fairweather, Abrol, ed. 2014. *Virtue Epistemology Naturalized: Bridges Between Virtue Epistemology and Philosophy of Science.* New York: Springer.

Floridi, Luciano, and Mariarosaria Taddeo, eds. 2014. *The Ethics of Information Warfare.* New York: Springer.

Foucault, Michel. 1995. *Discipline and Punish: The Birth of the Prison.* New York: Vintage Books.

Frede, Dorothea. 2013. "The Historic Decline of Virtue Ethics." In *The Cambridge Companion to Virtue Ethics*, edited by Daniel C. Russell, 124–48. Cambridge: Cambridge University Press.

Frost-Arnold, Karen. 2013. "Moral Trust and Scientific Collaboration." *Studies in History and Philosophy of Science Part A* 44, no. 3: 301–10.

Frost-Arnold, Karen. 2014. "Imposters, Tricksters, and Trustworthiness as an Epistemic Virtue." *Hypatia* 29, no. 4: 790–807.

Harris, Charles E. 2008. "The Good Engineer: Giving Virtue Its Due in Engineering Ethics." *Science and Engineering Ethics* 14, no. 2: 153–64.

Harrison, Peter. 2015. *The Territories of Science and Religion.* Chicago: University of Chicago Press.

Herzig, Rebecca M. 2005. *Suffering for Science: Reason and Sacrifice in Modern America.* New Brunswick, NJ: Rutgers University Press.

Hicks, Daniel J., and Thomas A. Stapleford. "The Virtues of Scientific Practice: MacIntyre, Virtue Ethics, and the Historiography of Science." *Isis: Journal of the History of Science Society* 107, no. 3 (September 7, 2016): 449–72. https://doi.org/10.1086/688346.

Hottois, Gilbert. 2006. "La technoscience: De l'origine du mot à ses usages actuels." *Recherche en soins infirmiers* 86, no. 3: 24–32.

Hull, David L. 1988. *Science as a Process: An Evolutionary Account of the Social and Conceptual Development of Science.* Chicago: University of Chicago Press.

Jones, Matthew L. 2006. *The Good Life in the Scientific Revolution: Descartes, Pascal, Leibniz, and the Cultivation of Virtue.* University of Chicago Press.

Kevles, Daniel J. 1977. *The Physicists: The History of a Scientific Community in Modern America.* New York: Knopf.

Marr, Alexander, Raphaele Garrod, Jose Ramon Marcaida, and Richard J. Oosterhoff. 2019. *Logodaedalus: Word Histories of Ingenuity in Early Modern Europe.* Pittsburgh: University of Pittsburgh Press.

Martin, Mike W. 2007. *Creativity: Ethics and Excellence in Science.* Lanham, MD: Lexington Books.

Martin, Mike W., and Roland Schinzinger. 2010. *Introduction to Engineering Ethics.* New York: McGraw-Hill.

Pennock, Robert T. 2019. *An Instinct for Truth: Curiosity and the Moral Character of Science*. Cambridge, MA: MIT Press.

Pickering, Andrew, ed. 1992. *Science as Practice and Culture*. Chicago: University of Chicago Press.

Polanyi, Michael. 1964. *Personal Knowledge: Towards a Post-Critical Philosophy*. New York: Harper & Row.

Russell, Daniel C. 2014. "Aristotle on Cultivating Virtue." In *Cultivating Virtue: Perspectives from Philosophy, Theology, and Psychology*, edited by Nancy E. Snow, 17–48. Oxford: Oxford University Press.

Schatzki, Theodore R., Karin Knorr-Cetina, and Eike von Savigny, eds. 2001. *The Practice Turn in Contemporary Theory*. New York: Routledge.

Scherz, Paul. 2019. *Science and Christian Ethics*. Cambridge: Cambridge University Press.

Shapin, Steven. 2009. *The Scientific Life: A Moral History of a Late Modern Vocation*. Chicago: University of Chicago Press.

Soler, Lena, ed. 2014. *Science After the Practice Turn in the Philosophy, History, and Social Studies of Science*. New York: Routledge.

Stovall, Preston. 2011. "Professional Virtue and Professional Self-Awareness: A Case Study in Engineering Ethics." *Science and Engineering Ethics* 17, no. 1: 109–32.

Stump, David J. 2007. "Pierre Duhem's Virtue Epistemology." *Studies in History and Philosophy of Science Part A* 38, no. 1: 149–59.

Vallor, Shannon. 2016. *Technology and the Virtues: A Philosophical Guide to a Future Worth Wanting*. New York: Oxford University Press.

Wang, Jessica. 2017. "'Broken Symmetry': Physics, Aesthetics, and Moral Virtue in Nuclear Age America." In *Epistemic Virtues in the Sciences and the Humanities*, edited by Jeroen van Dongen and Herman Paul, 27–47. Cham, Switzerland: Springer.

Warwick, Andrew. 1998. "Exercising the Student Body: Mathematics and Athleticism in Victorian Cambridge." In *Science Incarnate: Historical Embodiments of Natural Knowledge*, edited by Christopher Lawrence and Steven Shapin, 288–326. Chicago: University of Chicago Press.

Warwick, Andrew, and David Kaiser. 2005. "Kuhn, Foucault, and the Power of Pedagogy." In *Pedagogy and the Practice of Science: Historical and Contemporary Perspectives*, edited by David Kaiser, 393–409. Cambridge, MA: MIT Press.

White, Paul. 2009. "Introduction: The Emotional Economy of Science." *Isis* 100, no. 4: 792–97.

Wittgenstein, Ludwig. 1963. *Philosophical Investigations*. Translated by G. E. M. Anscombe. Oxford: Blackwell.

PART I
VIRTUES IN SCIENCE

1

Sacrifice in Service to Truth

The Epistemic Virtues of Victorian British Science

Richard Bellon

George Peacock, Cambridge mathematician and dean of the Anglican cathedral at Ely, celebrated the ecumenical spirit of the British Association for the Advancement of Science in his presidential address to its fourteenth annual meeting in 1844. He praised the association's founders for conceiving that "men of different shades of political opinion or religious belief would rejoice in the opportunities which such Meetings would afford them of coming together, as it were, upon neutral ground." While he scorned treating religious opinions with indifference as a "spurious and false liberality," he stressed that toleration in the shared pursuits of natural truth was itself a manifestation of the courtesy enjoined by Christian faith (Peacock 1845, xxxiii; Morrell and Thackray 1981).

Peacock explicitly connected the association's religious identity with its members' Christian virtues in eulogies for two celebrated scientists. The life and work of the Quaker chemist John Dalton was rooted in his "long-continued and persevering thought." "Steady perseverance" likewise characterized the Anglican astronomer Francis Baily (Peacock 1845, xxxvii, xli). The perseverance of the Quaker chemist and the Anglican astronomer did not merely exemplify the qualities necessary for scientific investigation or create a basis for productive social intercourse. In the view of Peacock and his fellow leaders of the British Association, self-denying virtues permeated science with Christian character. Peacock's articulation of epistemic virtue directly followed the duties of Christian behavior drawn by William Paley in *Evidences of Christianity* (1794), a book that Peacock endorsed as compulsory study for every Cambridge undergraduate (Peacock 1841, 161; Clarke 1974, 129). Paley grounded *Evidences* on "the *character of Christ*." The story of the Garden of Gethsemane was central to Paley's understanding of that character. Christ prayed in agony to forgo the trials of crucifixion, but in the end he did not hesitate to drink from the cup God had given him. This act of self-denial defined the nature of Christianity. Peter, Paul, and the other founders of the early church accepted murderous harassment as the price of their faith. This holy resignation warranted the truth claims of Christianity because, said Paley, it strained credulity that the apostles would

Richard Bellon, *Sacrifice in Service to Truth* In: *Science, Technology, and Virtues*. Edited by: Emanuele Ratti and Thomas A. Stapleford, Oxford University Press. © Oxford University Press 2021. DOI: 10.1093/oso/9780190081713.003.0002 court "lives of fatigue, danger, and suffering" to spread falsehood (Paley [1794] 1838, 50, 123, 177–79, 196). He put the matter bluntly in a sermon: "Produce me an example of any one man, since the beginning of the world, voluntarily suffering death for what he knows to be false, and I give up the cause" (Paley 1838, 515).

The fortitude needed to accumulate observational and experimental results and the honesty required to judge them properly did not demand such grave sacrifices—neither, typically, did the modern practice of Christianity, for that matter—but these practices were still severe enough to demand the self-less habits of patience and humility that characterized Christ and his apostles. For this reason, scientific discovery was a moral process, not an isolated event. Scientists deployed a long list of words to imbue favored scientific research with moral authority—these included *ardent, arduous, careful, diligent, disinterested, humble, impartial, indefatigable, industrious, laborious, methodical, painstaking, patient, perseverant, scrupulous,* and *zealous.*[1]

This broad consensus on the epistemic virtues necessary to scientific investigation allowed—usually, if not always—leading Victorian scientists to collaborate despite diverse religious and political commitments. While in practice the British Association did not equally welcome all political and religious preferences, it nonetheless brought into its fold individuals with a wide variety of beliefs, including, after midcentury, avowed agnostics such as Thomas Huxley and John Tyndall.

If a shared vocabulary of epistemic virtue provided Victorian scientists with common ground, it could also act as an accelerant to their ideological conflicts. The conviction that out of virtue came truth had a potent corollary. Vice begat error. Arrogance and pride left an investigator dangerously fond of his own speculations. The problem was not speculation itself but an aversion to the tedious investigation that would test, refine, and frequently invalidate leaps of imagination (Whewell 1837, 1:411–12; Herschel 1830, 190–91, 204; Herschel 1857, 514–15). A thirst for applause and fortune led to the betrayal of truth whenever it conflicted with popular fashion. Disagreements over ideas often provoked accusations of immoral behavior. Accusation of error could escalate into denunciation of sin.

Framing scientific investigation in moral as well as intellectual terms could degenerate into a vicious circularity. There was the constant temptation to assume that appealing ideas emerged from virtue, which in turn validated their truth, or that an unwelcome conclusion must be the fruit of vice. The Victorians often weaponized the association of proper scientific investigation with virtuous self-abrogation. If the production of reliable natural knowledge required a dutiful subordination of individual interest, desire, and comfort to arduous and impartial investigation, then it followed that greed, arrogance, and laziness would invariably distort the understanding and produce dangerous error.

The rhetorical embrace of self-denial by Victorian scientists was not simply a façade for ideological, economic, or social interests, however. All leading scientists did succumb to some degree to self-interestedness, self-delusion, and self-congratulation. But for all their human fallibility, they were typically sincere and self-aware. They emphasized human imperfection—including their own— in mental processes as well as physical perceptions. The goal of science's methods and social organization was directly analogous to the creation and use of new instruments, which aimed to compensate for human imperfection. The physicist James Clerk Maxwell considered "pains and patience" as integral to the function of a laboratory as its apparatus (Maxwell 1879, 311–12).

Preoccupations with epistemic virtue have permeated scholarly thought in many different times and places, across intellectual and cultural traditions, and in different religious and political communities (Van Dongen and Paul 2017; Bellon 2015). This essay concentrates on one of the epicenters of Victorian science, the University of Cambridge. It offers a particularly vivid case study of how considerations of epistemic virtue decisively influenced the methods and the intellectual content of science.

Virtue and Science at the University of Cambridge

In 1876, Henry Sidgwick sketched the revitalization of philosophy at the University of Cambridge over the previous half century. John Herschel's *A Preliminary Discourse on the Study of Natural Philosophy* (1830) had marked a watershed in this renewal by inspiring Cantabrigians of a mathematico-physical bent to approach philosophy with a new seriousness and creativity. Adam Sedgwick's *Discourse on the Studies of the University* (1833) had provided another jolt of fresh thought. These two discourses were prelude to the monumental work of William Whewell, particularly his three-volume *History of the Inductive Sciences* (1837) and his two-volume *Philosophy of the Inductive Sciences* (1840). These "elaborate investigations of the methods of modern science" became load-bearing pillars in a new edifice of teaching and investigation at Cambridge. "It is to Whewell more than to any other single man that the revival of Philosophy in Cambridge is to be attributed," Sidgwick wrote (Sidgwick 1876, 238–42; on practical virtue and self-control in Sidgwick's philosophy, see Schneewind 1977 and Crisp 2015). This was a philosophy rooted in a union between virtue and science, articulated by men whose influence dominated Cambridge and profoundly influenced Victorian British science more generally.

John Herschel: Self-Restraint in the School
of Just Subordination

As an undergraduate, Herschel led a successful effort to reform the university's
moribund mathematics education (Snyder 2011). In the years after graduation,
he accumulated triumphs in mathematics, astronomy, chemistry, acoustics, and
optics. He won the Royal Society's Copley Medal in 1821, before his thirtieth
birthday, for his mathematics. In awarding the medal, Humphry Davy celebrated
the young man for embodying Virgil's motto "Virtutem extendere factis"—valor
by deeds (Davy 1827, 3–10).

Thousands absorbed the lessons of Herschel's *Preliminary Discourse* during
the course of the nineteenth century, including Charles Darwin and John
Stuart Mill. For decades journalists and editors pillaged it for pithy aphorisms
and morally uplifting passages to gild their books and articles. As James Secord
demonstrates, the *Preliminary Discourse* was far more than a treatise on scientific research. Herschel grounded the specific skills necessary for science in the
higher obligations of "good method and appropriate modes of thinking," Secord
shows; the *Preliminary Discourse* was a "conduct manual" dedicated to the principles of "moral probity, truth-telling, and religious faith" (Secord 2014, 80–106).

The reader of the *Preliminary Discourse* learned that science could progress
only under a self-denying "determination to stand and fall by the result of a direct appeal to facts in the first instance, and of strict logical deduction from them
afterwards." While he insisted on the necessity of theoretical speculation, this
liberty "is not like the wild license of the slave broke loose from his fetters, but
rather like that of the freeman who has learned the lessons of self-restraint in
the school of just subordination." A long passage from Cicero's *De Officiis* served
as the book's motto: "Before all other things, man is distinguished by his pursuit and investigation of TRUTH." From this earnest longing "spring greatness of
mind and contempt of worldly advantages and troubles" (Herschel 1830, vii, 79,
190–91).

Herschel's call to radically disinterested investigation was not a platitude. He
spoke with genuine authority on patient investigation and humble submission to
the tests of observation and experiment because he practiced these values with
fidelity. He learned early, particularly in astronomical observation, that reliable
science demanded tedious, exhausting, and often fruitless effort, pursued day
after week after month.

This is not to say that Herschel always remained unruffled. On the contrary,
his frustration often boiled over. He bemoaned in a letter to his wife that he saw
"two stars last night, and sat up till two waiting for them. Ditto the night before.
Sick of star-gazing—mean to break the telescope and melt the mirrors." After
this burst of self-pity, he announced the Cicero quotation he had found for the

motto of the *Preliminary Discourse* (Buttmann 1970, 49–51; Crowe, Dyck, and Kevin 1998, letter 2192). This juxtaposition captures the role of epistemic virtue in Herschel's life and career. Cicero's noble sentiments did not sit in tension with his irritation at the tedium of observation. Nor did Cicero adorn the *Preliminary Discourse* merely as a splash of cultured moral uplift. In Herschel's life, the principle of sacrifice in service to truth dissipated the temptation to abandon his telescopes for less annoying ways to spend his nights. The *Preliminary Discourse* sought to inspire readers with a similar resilience in the face of the inevitable vexations and inconveniences of research.

Father Adam's Crusade Against Humbug

Sedgwick's *Discourse on the Studies of the University* originated in a two-hour sermon preached in the Trinity College Chapel in December 1832 (Bellon 2012, 937–58). Sedgwick expanded the text into a book the following year. The fourth edition appeared in 1835. Sedgwick added an additional five hundred pages to a fifth and final edition in 1850. The *Discourse* was animated by Sedgwick's conviction that "the moral capacities of man must not be left out of account in any part of intellectual discipline." He developed this theme most fully in his unification of "moral and religious habits of thought" with the "severe" studies of natural phenomena, which were "on the whole favourable to self control." The study of nature enhanced intellectual powers, but "strange must be our condition of self-government and tortuous our habits of thought, if such studies be allowed to co-exist with self-love and arrogance and intellectual pride." He prepared the book chiefly for the young men of Cambridge, but hoped that it was "not altogether unfitting to other ears" (Sedgwick 1833, 7–12).

The *Discourse* galvanized Sedgwick's friends and allies. His acolyte Joseph Beete Jukes reveled at "Father Adam" speaking in the spirit of Thomas Carlyle to drive "a good broad harpoon deep into the sweltering sides of the floating carcass of Humbug" (Browne 1871, 447). Yet Sedgwick's uncompromising moral convictions and intellectual commitments remained scattershot. John Stuart Mill took advantage of this lack of philosophical coherence in an unsparing review of the *Discourse*, which blasted Sedgwick's educational principles and anti-utilitarianism. If general opinion put such a shoddy thinker at the forefront of Cambridge minds, he said, then the university had genuine cause for shame. Notably, however, Mill never disputed Sedgwick's definitions of epistemic virtue, although he did declare them "trite commonplaces" whose origin Sedgwick misunderstood (Mill 1835, 100–101, 112, 134). A few years earlier, in reflections on Herschel's *Preliminary Discourse*, Mill celebrated physical science not for its conclusions but for "the process by which it obtains them." It offered the sole example "of a vast body of connected truth, gradually elicited by patient and earnest investigation" (Mill [1831] 1986, 285). Mill's anger flared over Sedgwick's attempt to claim exclusive ownership of epistemic virtue on behalf of anti-utilitarianism.

Mill subsequently turned his back on Sedgwick and trained his fire on a philosophically weightier Cambridge opponent (Snyder 2006, 98, 229–30). That opponent—Sedgwick's close friend Whewell—offered Sedgwick a backhanded compliment when he dedicated the *Philosophy of the Inductive Sciences* to him. "I have little doubt," Whewell wrote, "that if your life had not been absorbed in struggling with many of the most difficult problems of a difficult science, you would have been my fellow-labourer or master in the work which I have here undertaken" (Whewell 1840, 1:iii–iv). Be that as it may, Whewell created a coherent philosophical strategy to guide the tactical use of the humbug-skewering harpoon.

William Whewell: The History and Philosophy of Scientific Virtue

Herschel observed in the *Preliminary Discourse* that history offered a valuable means to understand the general principles of scientific discovery, but the scope of his present work prevented him from offering any systematic historical analysis. "We are not, however, without a hope that this great desideratum in science will, ere long, be supplied from a quarter every way calculated to do it justice," he reported (Herschel 1830, 219–20). Whewell fulfilled this hope in 1837 with his *History of the Inductive Sciences*. This monumental work argued that the epistemic virtue of self-denial was an absolute necessity for scientific discovery.

Whewell's father was a Lancaster master carpenter who scrimped to send his son to Cambridge. He died shortly before his son graduated laureled with high academic honors (Snyder 2011). Whewell reassured himself that his father would have greeted his academic success as "a sufficient recompense for the unwearied exertions of a laborious life" (Stair Douglas 1881, 10, 23–24). He never forgot that his achievements were possible only because of his father's quiet life of rectitude and self-sacrifice.

Unlike Herschel, Whewell never left Cambridge. A Trinity College fellowship followed his graduation in 1817, and from there he climbed relentlessly into academic positions of ever-greater authority and prestige. He was a dominant voice in the British Association and a fixture at the Royal, Astronomical, and Geological Societies. His career reached its pinnacle in 1841 when he was appointed to the mastership of Trinity College, making him the most influential man in Cambridge.

Whewell was no stranger to sustained scientific research. His extensive work on the theory of tides won a Royal Medal from the Royal Society in 1837 (Reidy 2008). By this time, however, he had already decided to make the methodical reform of philosophy his "great object" (Todhunter 1876, 2:233–36). Herschel continued his exhausting astronomical observations. Sedgwick maintained his grueling geological work in the field and museum. By the time of the publication of his *History*, Whewell decided that his contribution of "weariness" to the progress of science would consist "in writing a multitude of books" (Whewell 1838, iii).

Whewell made an important statement about history, scientific discovery, and virtue in an influential address to the British Association in 1833. In reflections on "what I may call the morals of science," he defended science against the claims that scientific progress tended to make "men confident and contemptuous, vain and proud." The history of science decisively falsified this criticism, he said. Great discoverers were overwhelmingly "sober and modest" because such qualities were necessarily to the "patience and labour" of investigation. "Knowledge, like wealth, is not likely to make us proud or vain, except when it comes suddenly and unearned," he reassured his audience, "and in such a case, it is little to be hoped that we shall use well, or increase, our ill-understood possession" (Whewell 1834, xxiii–xxv). He started writing the *History* six months later. This principle that hard-won knowledge depended on virtue became its beating heart (Cantor 1991, 69).

In the years following publication of the *History*, Whewell developed a theory that accounted for the interactions of hypotheses, mind, creativity, genius, social organization, rational ideas, and empirical evidence in scientific investigation. He developed a theory of induction that privileged fundamental ideas over empiricism, which provoked disagreement both respectful (as from Herschel) and vituperative (as from Mill) (Fisch and Schaffer 1991; Yeo 1993; Snyder 2006). As Whewell explained to a friend, the purpose of his "grand projects" in history and philosophy was to imbue "a right and wholesome turn to men's minds" (Todhunter 1876, 2:175).

Whewell dedicated the *History* to Herschel "in affectionate admiration of your moral and social, as well as intellectual excellencies." The three volumes were filled with heartwarming stories of individual moral excellence. Isaac Newton served as the ultimate exemplar for "those who love to think that great talents are naturally associated with virtue." Whewell's Newton combined "the utmost devotion of thought, energy of effort, and steadiness of will" with "candid and humble, mild and good" personal conduct (Whewell 1837, 1:v–vi, 2:183–87; Higgitt 2007; Bellon 2014).

Whewell noted that Newton's triumph was not recognized uniformly across Europe. France in particular remained in thrall to René Descartes long after the

English adopted Newton's *Principia*. Whewell attacked Descartes in tellingly personal terms. He reproached the French philosopher for "great conceit and great shallowness" in turning his back on mathematicians who were "patiently labouring to bring the mechanical problem of the universe into its most distinct form." No part of Descartes's philosophy remained untainted by his "rash and cowardly" character. These moral failings were rooted in his thirst for deductive speculation, which allowed him to reach putatively certain conclusions without the irksome observations and experiments of Newton's inductive approach. For Whewell, England's glory rested not just in having produced Newton but in immediately honoring his transcendent generalizations. France, on the other hand, suffered under an oppressive religious and political establishment that explained its blighted response to Newton's discoveries (Whewell 1837, 2: 132–37, 190–205).

Whewell even more bluntly invoked national and religious characteristics in his account of the condemnation of Galileo Galilei. In the *Preliminary Discourse*, Herschel had cast the story in conventional terms of "persecutions, . . . perseverance and suffering" (Herschel 1830, 113). Whewell recounted it differently. In his telling, Galileo lacked the courage to accept true martyrdom in defense of philosophical and religious truth. He insincerely renounced his beliefs, and the authorities cynically pretended to believe him. Far from enduring the dungeons of the Inquisition, Galileo suffered nothing more arduous than comfortable house arrest. The whole sorry affair revealed "Italian traits of character," which, Whewell concluded, were so unlike "the vigorous independent habits of thought of Germany and England" (Whewell 1837, 1: 399–401).

The story of Galileo, like that of Descartes after him, offered a clear lesson—the temper of nations and ages, no less than the conduct of individuals, bore the stamps of good and evil. Whewell later moderated this harsh judgment. In the *Philosophy of the Inductive Sciences* (1840), he credited "the learned in Italy" for the revitalization of philosophy and acknowledged Galileo as a practical reformer who introduced a salutary caution and prudence into science (Whewell 1840, 2:154–57, 340–42, 366–70, 379–83). In the 1847 second edition of the *History*, Whewell excised the passage about Italian servility, although he refused to back down from his conclusion that Galileo was neither a genuine martyr nor deserving of "our unconditional admiration" (Whewell 1847, 1:418–26, 461–66).

Whewell's softening on Galileo and the Italian "character" responded to contemporary criticism. But it also reflected his conviction that the battle against intellectual darkness and barren dogmatism was internecine. He recognized modern England—no less than Galileo's Italy—as beset on all sides by dangerous philosophical and religious errors. Herschel's *Preliminary Discourse* and Sedgwick's *Discourse* both made earnest pleas for the unification of science and religion around shared standards of virtuous conduct. Whewell provided the dense and systematic historical and philosophical case that men of goodwill must join the battle against self-gratifying philosophical systems.

Virtue as Shield and Sword

The Cambridge scientific tradition was organized around the cultivation and organization of self-denial. Virtue provided the university's scientific gentlemen with both a shield and a weapon against the spate of opponents.

In the 1830s, one of the most dangerous threats was the backlash to geology from a heterogeneous group of "scriptural geologists" who attempted to harmonize their (often mutually incompatible) readings of the Bible with the physical evidence of geology. The 1840s were marked by attempts to undermine popular interest in evolution, particularly as promoted by the anonymously authored bestseller *Vestiges of the Natural History of Creation* (1844). Both the mysterious "Mr. Vestiges" and the scriptural geologists lacked relevant research experience, which made them susceptible to attacks on their virtue.

By the middle of the 1850s the scientific elite had reason to feel confident that they had neutralized, if not entirely vanquished, these threats. The association of mainstream British science with Christianity, based on the shared commitment to virtues of humble and patient self-control, appeared ascendant. It was not. Herschel, Whewell, and Sedgwick in their old age watched scientific theism lose its exclusive ownership of epistemic virtue in the organizations they had built, including the British Association and the Geological Society. Their religious principles were not overthrown from outside. On the contrary, the most important catalyst was the work of a close disciple with an unbreakable dedication to the epistemic values and traditions of Cambridge science. Darwin's theory of evolution by natural selection did not challenge his Cambridge mentors' ideals for the virtuous practice of science. Rather, he showed that these ideals did not have to flatter their cherished religious assumptions.

The Hammer-Bearing Philosophers Confront Scriptural Geology

By the 1820s, the fellows of the elite Geological Society had developed theories about the history of the earth that were increasingly unbound from Genesis. Sedgwick and Whewell both served as president of the society. Herschel followed geology knowledgeably and made reliable field observations. At society meetings he sat in honor among the leadership even if he did not join it (Lyell 1881, 1:372).

Whewell's *Essay on Mineralogical Classification and Nomenclature* (1828) framed mineralogy exclusively as the quest for relevant natural laws (Whewell 1828, i). He did not touch on any religious implications. This is not because they did not exist. As Whewell understood, the science of mineralogy owed a decisive debt to men such as the eighteenth-century German naturalist Abraham Gottlob Werner, who had extended the age of the earth into the hundreds of thousands of years (Whewell 1837, 3:224–26; Rudwick 2005, 59–62, 125–26). This disregard of a literal reading of Genesis had raised hackles from the first. In nineteenth-century Britain, geology's search for general laws engulfed it in particularly acute controversy (O'Connor 2007a, 133–45; 2007b, 357–403).

In their reflections on scientific method Herschel, Whewell, and Sedgwick all wove in the principle that science should never be subject to scriptural tests (with a carve-out for conclusions that touched on the moral condition of humanity). Herschel's *Preliminary Discourse* explicitly denounced "such reasoners as would make all nature bend to their narrow interpretations of obscure and difficult passages in the sacred writings." The modern world, he observed, had turned its back on such misbehavior (Herschel 1830, 9–10). Elsewhere, he celebrated science for its ability "to lead us up, by legitimate induction, to its Author." This morally enriching contemplation of nature depended on its freedom. He insisted that "we must be careful to raise up no self-created phantasms of our own minds, interposing an impassable barrier to further progress, and cutting off the chain of connexion by a stern *ne plus ultra*" (Herschel 1848, 178).

Whewell condemned atheists as "odious" and considered deists as hardly better (Whewell 1845, 2:127–28). But he also insisted that any attempt to frame physical sciences such as astronomy or geology according to scripture "is a perversion of the purpose of a divine communication, and cannot lead to any physical truth" (Whewell 1837, 1:400–403, 3:601–2; 1832, 117). Sedgwick cautioned against misinterpreting figurative passages in scripture. "These writings deal not in logical distinctions or rigid definitions," he argued. "They were addressed to the heart and understanding, in popular forms of speech such as men could readily comprehend" (Sedgwick 1833, 104–5).

Sedgwick's presidential addresses in 1830 and 1831 marked key moments in the debate over the religious implications of geology. In the first, he lamented the recent rise of biblical literalist cosmology. He accused these speculators of miscomprehending the basic truth that "slow and toilsome induction . . . [is] the only path which leads to physical truth." He contemptuously waved away accusations that geologists undermined religious truth. "No opinion can be heretical but that which is not true," he insisted, adding that "we have nothing to fear from the results of our inquiries, provided they be followed in the laborious but secure road of honest induction" (Sedgwick [1830] 1826–32, 207–9).

In the following year's address he carefully associated geological processes with the wisdom of God and reassured his audience that their science was "connected with the loftiest moral speculations." He accentuated this lesson with a dramatic confession: he acknowledged that he had once attempted to understand geological phenomena in light of the Noachian deluge. His dispassionate examination of the material records of Earth's history, along with the work of his equally dedicated colleagues, falsified the theory of a universal flood. He did not recant flood geology merely as error. Rather, as he explained to the assembled geological fellows, he had fallen prey to the self-indulgent passion of a "philosophic heresy" (Sedgwick [1831] 1826–32, 313–16). In these two addresses, Sedgwick demonstrated to the satisfaction of his audience of practitioners that scriptural geology was more than bad science. It was religiously heretical.

The battle between the geologists and their detractors ultimately hinged on the relative virtue of the disputants' rival approaches to natural phenomena. The elite gentlemen of the Geological Society could—and certainly did—mock their rivals' elementary misunderstanding of the accumulated physical evidence. This appeal to facts could carry the attack only so far, however. The scriptural geologists impeached secular reasoning by appealing to the higher authority of divine revelation.

In 1834, for example, the Rev. Henry Cole scorned the "moral impiety" of geology's dismissal of God's sacred testimony. "But alas! what a fool (in a modern scientific's estimation) was Adam, compared with a Geologist and Naturalist of the fifty-ninth century of the world! . . . What were the philosophy and astronomy of Adam, then," he taunted, "compared with those of the REV. ADAM SEDGWICK!" Cole for decades pressed for the absolute supremacy of scripture—or at least of his reading of it—over all forms of secular reasoning (Cole 1834, 59). Geologists such as Sedgwick addressed this challenge by highlighting the virtue of their practice—and the vice of their opponents'. By framing the controversy as a matter of proper conduct, the geologists attempted to command the moral and religious high ground.

Whewell reassured the conservative Anglican readers of the British Critic in 1831 that they could serenely leave geology in the hands of the "hammer-bearing philosophers." The English school of geology, he explained, was dominated by "a body of active, inquiring men" who eagerly sacrificed their physical and mental comfort in pursuit of truth. He noted that until relatively recently these paragons of philosophical virtue had abstained from theory-making with "self-denial and temperance." The recent burst of system-building by Charles Lyell and a few others was not the negation of this spirit of self-command, Whewell insisted, but its logical culmination. Geologists had diligently accumulated such a rich storehouse of evidence that they could start to build—and revise and contest—general theories of the globe's history. In short, they had earned their intellectual liberty and "the time is now come when no other intelligent persons look with suspicion on this class of speculations" (Whewell 1831, 180–82, 184, 205–6).

Sedgwick, operating from this perspective, took furious glee in attacking scriptural geologists for religious apostasy. He accused them, accurately, of lacking even the most rudimentary familiarity with geological work in field or museum and then framed this lack of expertise as damnable moral failure. Because all genuine geologists "followed in the laborious but secure road of honest induction," there was no cause to fear that their physical discoveries could truly conflict with religious truth (Sedgwick [1830] 1826–32, 207). On the other hand, the ignorant purveyors of scriptural geology attempted to make "a pattern of our own . . . by shifting and shuffling the solid strata of the earth, and then dealing them out in such a way as to play the game of an ignorant or dishonest hypothesis." No wonder, he said, that such efforts, riddled as they were with the vice of self-sufficiency, would produce only "mischievous nonsense" and "idle dreams" (Sedgwick 1833, 105–9).

Sedgwick make it clear that he would never step back from the "literal application" of scripture; he just insisted that biblical lessons had nothing to do with attempts to dictate scientific theory. Scripture taught that everyone must extinguish their egotism and submit to the divine will. Scriptural geologists failed in this *literal* duty. Sedgwick cast them as inhumane men who thirsted to crush "every truth not hatched among their own conceits, and confined within the narrow fences of their own ignorance." In Sedgwick's telling, scriptural geology turned its back on these "simple and affecting lessons of Christianity." This left it little more than paganism disguised under a tawdry cloak of distorted scriptural doctrine (Sedgwick 1833, 10–12, 76–77, 106).

Counterattacks against scriptural geology were not simply ad hoc defensive measures. They drew upon—and in turn sharpened the articulation of—a broader understanding of the relationship between science, religion, and moral behavior. The scriptural geologists, by cobbling together systems from books and imagination alone, had severed the production of scientific *belief* from its material *practice*. There were good workaday reasons for condemning such proceedings. An original investigator needed a tactile acquaintance with scientific evidence in order to understand, apply, and synthesize it reliably. But the eagerness to theorize in the absence of firsthand expertise (or even an interest in acquiring it) demonstrated a comprehensive moral failure. For the gentlemanly scientific elite, both the study of the natural world and the practice of Christianity depended on patterns of behavior and not merely on creed. From their perspective, bending scientific theory to the reading of the sacred writings represented unchristian behavior.

Mr. Vestiges's Frenzied Dream

Sedgwick did not just condemn scriptural geology in the first edition of his *Discourse*. He also excoriated another class of villains, those who "would rid themselves of a prescient first cause, by trying to resolve all phenomena into a succession of constant material actions, ascending into an eternity of past time." Among this class he singled out evolutionists. Their materialist theory was "no better than a phrensied dream" (Sedgwick 1833, 23). At that time no popular exposition of evolutionary theory had appeared in English. This was no longer true when Sedgwick published the fifth edition of the *Discourse* in 1850; much of the additional hundreds of pages attacked the "bad eminence" of *Vestiges of the Natural History of Creation*. He condemned the book as "shallow, one-sided, and inadequate." Its anonymous author, Sedgwick further charged, "has taken up a rash hypothesis, and made up his mind to defend it at all cost." This misbehavior explained the book's philosophical failures and its religious apostasy as two sides of the same debased coin. In its self-sufficiency and shameful ignorance, *Vestiges* was the materialist evil twin of scriptural geology (Sedgwick 1850, cl–cliii, 111–12, 247).

Herschel and Whewell also ruthlessly attacked *Vestiges*'s sweeping and vivacious evolutionary cosmology (Secord 2003). The gentlemanly leaders of British science deployed the same strategy that had worked so effectively against biblical literalists. *Vestiges* was attacked not merely for its errors but as a dangerous product of delinquency by its secretive author.

"The question lies," Whewell scolded in his *Indications of the Creator* (1845), "not between two philosophical theories, but between one philosophical theory, and a broken and obscure view of nature." This poisonous obscurity, he continued, was the direct fruit of the author's "headlong haste and baseless confidence" (Whewell 1846, 22–25, 29). Sedgwick detected the cloying perfume of effeminacy. Even if the author of *Vestiges* was not actually a woman, as Sedgwick claimed to have initially suspected in his 1845 attack on the book in the *Edinburgh Review*, its "rash and incongruous conclusions" singularly lacked hearty masculine virtue. "The ascent up the hill of science is rugged and thorny, and ill-fitted for the drapery of a petticoat," he declared, "and ways must be passed over which are toilsome to the body, and sometimes loathsome to the senses" (Sedgwick 1845, 3–4; on the relationship between virtue and gender in Victorian science, see Richards 2017; Ellis 2017; Bellon 2018). Mr. Vestiges's apparent lack of scientific experience was not just intellectually disqualifying but also morally disqualifying, as it had been for the scriptural geologists before him.

The leaders of the British Association rallied respectable scientific opinion against *Vestiges*. Herschel's 1845 presidential address to the association condemned the "propensity which is beginning to prevail widely, and, I fear, balefully, over large departments of our philosophy, the propensity to crude and over-hasty generalization." This mischievous tendency contrasted with Herschel's repeated praise of the "patient and persevering" discipline that characterized the civilizing practice of physics, astronomy, mathematics, and natural history (Herschel 1846, xxvii–xxviii, xxxvi, xliv).

By the early 1850s, neither scriptural geology nor evolutionary theory appeared to have much of a future. In 1852, Edward Forbes reassured the readers of the radical *Westminster Review* that even though "vague general distrust" of modern geology remained common in "respectable country society" and evangelical congregations, "geology is in the ascendant" (Forbes 1852, 67–68, 75–77). Two years later, the young and professionally struggling Thomas Huxley eviscerated the tenth edition of *Vestiges* for its "total absence of that careful research and fair representation of both sides of a question." He tellingly contrasted this viciously slapdash conduct to Sedgwick's sobriety. The geologist "has made truth the search of his life, and knows the difficulties of the road and the stern practical discipline required for success" (Huxley 1854, 438).

While respectable scientists had indeed deeply buried biblical literalist geology, the consensus against evolution was (in retrospect) broad but shallow. Huxley deplored the philosophical and methodological failures of *Vestiges* but fretted not at all about its alleged denial of a divine First Cause. The eminent London comparative anatomist Richard Owen began to murmur about finding secondary causes for the origin of new species, even if he quickly retreated in the face of Sedgwick's thinly veiled warnings about this dangerous trajectory (Rupke 1994, 188–204). The geologist Charles Lyell sensed that the intellectual and cultural winds had started to shift, and when briefed on a friend's private evolutionary theory in 1856, he encouraged its author to publish to avoid being forestalled (Browne 1995, 541).

Darwin resisted Lyell's prescient advice. In 1858, he received a letter from Alfred Russel Wallace setting out a theory remarkably similar to his own. To preserve his priority, he jointly announced his theory with Wallace's at the Linnean Society in 1858 (Browne 2002, 14–39). The following year he elaborated his theory of evolution by natural selection in *On the Origin of Species*—and sparked the greatest conflagrations over virtue and science of the Victorian period.

Darwin Conquers the Tram-Road of Physical Truth

Sedgwick accused Darwin of abandoning "that tram-road of all solid physical truth—the true method of induction" in a letter written the same day that the *Origin of Species* went on sale to the public. (Darwin had sent Sedgwick a presentation copy.) Sedgwick tried to soften the blows by reassuring Darwin that he considered him "a good tempered & truth loving man" and that he meant his criticisms "in a spirit of brotherly love" (Burkhardt et al. 1985–, 7:396–98; Bellon 2012). Darwin felt wounded if not surprised at Sedgwick's "rabid" reaction (Burkhardt et al. 1985–, 8:134, 140).

Sedgwick's private letter previewed the torrent of public moral condemnation about to break over Darwin's head. One hostile review after another reproached Darwin for betraying the virtues of proper scientific research. Owen in the *Edinburgh Review* indicted Darwin for spurning "the patient and honest study and comparison of plants and animals" for the vain pleasure of "unregulated fancies of dreamy speculation" (Owen 1860, 512–13). In the *Quarterly Review*, Bishop Samuel Wilberforce condemned Darwin for "utterly dishonourable" conduct in defiling "the sober, patient, philosophical courage of our home philosophy" (Wilberforce 1860, 250, 263). These and similar attacks had enough initial success that Owen predicted that the *Origin* would be forgotten within a decade. Darwin, pummeled into despondency by hostile reviews, sometimes feared that Owen might prove right (Burkhardt et al. 1985–, 8:272).

In the end, however, the early campaigns against the *Origin*'s legitimacy— and Darwin's reputation—won only pyrrhic victories. Darwin had the ability to demonstrate the usefulness of evolution and natural selection as guides to original investigation. He did so most prominently in a series of impeccably precise studies that used evolution to revolutionize the scientific understanding of the reproductive biology of flowering plants. Darwin's post-*Origin* botany contributed empirical confirmation of evolution and, even more importantly, demonstrated that it could serve as a productive research tool. And it thoroughly rehabilitated his character (Bellon 2003, 2009, 2011, 2013; Tabb 2016; Endersby 2016, 81–105).

The famously sober botanist George Bentham exhorted the fellows of the Linnean Society in his 1862 presidential address to take "a leaf out of Mr. Darwin's book" in conducting their own investigations (Bentham 1862, lxxxiii). The same year, Oxford's elderly professor of botany Charles Daubeny encouraged an audience at the British Association to carefully study Darwin's recent book on orchid fertilization. It "was important to [Darwin's] reputation," Daubeny said, "as it would dispel many notions which had been wrongly entertained with regard to the tendency of his writings" ("British Association" 1862, 2). In 1866, Darwin's entry in *Portraits of Men of Eminence in Literature, Science, and Art* acclaimed him as an exemplar of scientific virtue: "Everywhere [in Darwin's work] we discover the same painstaking experimental investigation, the same close and long-continued observation; and also everywhere we discover that high power of drawing with clearness and simplicity his deductions from his well-established facts, which distinguishes the true *Philosopher*" (Walford 1863–68, 5:52)

The triumph of Darwin's theory unhitched the yoke between Christianity, science, and virtue that had defined the ethos of gentlemanly British science. The ecumenicalism that Peacock had celebrated in his presidential address to the British Association did not stretch far enough in 1844 to encompass anything like Huxley's agnosticism. A quarter century later, after the *Origin*, it did. The admiration Huxley expressed for Sedgwick's character in his 1854 *Vestiges* review might have been calculated, but it was not insincere. In the post-*Origin* years he publicly insisted that such epistemic virtue existed *in spite* of Christianity, not because of it. Science, he proclaimed to all who would listen, more efficiently cultivated virtue when unshackled from the supernaturalism, superstition, and dogmatism inherent in religious belief. When Huxley took his turn as British Association president in 1870, he identified the association's "justification and great glory" in its spirit of "diligent, patient, loving study" (Huxley 1871, lxxxix).

Conclusion

The conviction that all people must pay homage to the truth raised the central question of nineteenth-century British thought: how do we recognize what *is* the truth? This is an especially difficult problem given the conviction that man is a fallible creature with fallible perception. History—most notably that written by Whewell—showed that there were always disagreements over truth. Concepts of religious, metaphysical, or scientific truth did not remain stable over time. An eternal set of virtues—patience, humility, diligence, self-control—provided moral stability to science even as it relentlessly advanced new theories of physical reality.

This is why the scientific establishment, which coalesced around the annual meetings of the British Association, organized their individual and collective practices around ideals of self-denial. They disagreed about a great deal, sometimes raucously, about what to think. But as Darwin explained to his wife, Emma, to assuage her doubts about his lack of religious faith, "luckily there were no doubts as to how one ought to act" (Burkhardt et al. 1985–, 2:172).

Note

1. These examples come from Herschel 1857, 57, 88–89, 99, 111, 122–23, 144, 146, 154–55, 164–66, 170–71, 178–79, 262–63, 314, 26, 354–55, 385, 388, 466, 468–69, 472–73, 478, 479, 492, 498–501, 505–7, 511, 514–16, 552.

References

Bellon, Richard. 2003. "The Great Question in Agitation: George Bentham and the Origin of Species." *Archives of Natural History* 30: 282–97.

Bellon, Richard. 2009. "Charles Darwin Solves the 'Riddle of the Flower'; or, Why Don't Historians of Biology Know About the Birds and Bees?" *History of Science* 47: 373–406.

Bellon, Richard. 2011. "Inspiration in the Harness of Daily Labor: Darwin, Botany and the Triumph of Evolution, 1859–1868." *Isis* 102: 392–420.

Bellon, Richard. 2012. "The Moral Dignity of Inductive Method and the Reconciliation of Science and Faith in Adam Sedgwick's *Discourse*." *Science & Education* 21: 937–58.

Bellon, Richard. 2013. "Darwin's Evolutionary Botany." In *The Cambridge Encyclopedia of Darwin and Evolutionary Thought*, edited by Michael Ruse, 131–38. Cambridge: Cambridge University Press.

Bellon, Richard. 2014. "There Is Grandeur in This View of Newton: Charles Darwin, Isaac Newton and Victorian Conceptions of Scientific Virtue." *Endeavour* 38: 222–34.

Bellon, Richard. 2015. *A Sincere and Teachable Heart: Self-Denying Virtue in British Intellectual Life, 1736–1859*. Leiden: Brill.

Bellon, Richard. 2018. "Emotional Comfort and Theoretical Necessity: Sex and Gender in the Age of Darwin." *Historical Studies in the Natural Sciences* 48: 246–57.

Bentham, George. 1862. "Presidential Address Read at the Anniversary Meeting of the Linnean Society on Saturday, May 24, 1862." *Journal of the Proceedings of the Linnean Society (Zoology)* 6: lxvi–lxxxiii.

"The British Association for the Advancement of Science." 1862. *Daily News* no. 5117 (October 3, 1862), 2.

Browne, C. A., ed. 1871. *Letters and Extracts from the Addresses and Occasional Writings of J. Beete Jukes*. London: Chapman and Hall.

Browne, Janet. 1995. *Charles Darwin: Voyaging*. Princeton: Princeton University Press.

Browne, Janet. 2002. *Charles Darwin: The Power of Place*. Princeton: Princeton University Press.

Burkhardt, Frederick et al., eds. 1985–. *The Correspondence of Charles Darwin*. 26 vols. Cambridge: Cambridge University Press.

Buttmann, Günther. 1970. *The Shadow of the Telescope: A Biography of John Herschel*. Translated by B. E. J. Pagel. Edited with an introduction by David S. Evans. New York: Charles Scribner's Sons.

Chambers, Robert. 1844. *Vestiges of the Natural History of Creation*. London: John Churchill.

Cantor, Geoffrey N. 1991. "Between Rationalism and Romanticism: Whewell's Historiography of the Inductive Sciences." In *William Whewell: A Composite Portrait*, edited by Menachem Fisch and Simon Schaffer, 67–86. Oxford: Clarendon Press.

Clarke, M. L. 1974. *Paley: Evidences for the Man*. London: SPCK.

Cole, Henry. 1834. *Popular Geology Subversive of Divine Revelation! A Letter to the Rev. Adam Sedgwick, Woodwardian Professor of Geology in the University of Cambridge, Being a Scriptural Refutation of the Geological Positions and Doctrines Promulgated in His Lately Published Commencement Sermon, Preached at the University of Cambridge, 1832*. London: Hatchard and Son.

Crisp, Roger. 2015. *The Cosmos of Duty: Henry Sidgwick's Methods of Ethics*. Oxford: Oxford University Press.

Crowe, Michael J., David R. Dyck, and James R. Kevin, eds. 1998. *Calendar of the Correspondence of Sir John Herschel*. Cambridge: Cambridge University Press.

Davy, Humphry. 1827. *Six Discourses Delivered Before the Royal Society at Their Anniversary Meetings on the Award of the Royal and Copley Medals*. London: John Murray.

Ellis, Heather. 2017. *Masculinity and Science in Britain, 1831–1918*. London: Palgrave Macmillan.

Endersby, Jim. 2016. *Orchids: A Cultural History*. Chicago: University of Chicago Press.

Fisch, Menachem, and Simon Schaffer, eds. 1991. *William Whewell: A Composite Portrait*. Oxford: Clarendon Press.

Forbes, Edward. 1852. "The Future of Geology." *Westminster Review* 58: 67–94.

Herschel, John. 1830. *A Preliminary Discourse on the Study of Natural Philosophy*. London: Longman, Rees, Orme, Brown, and Green.

Herschel, John. 1846. "Presidential Address." In *Report of the Fifteenth Meeting of the British Association for the Advancement of Science; Held at Cambridge in June, 1845*, xxvii–xliv. London: John Murray.

Herschel, John. 1848. "Humboldt's *Kosmos*." *Edinburgh Review* 87: 170–229.

Herschel, John. 1857. *Essays from the Edinburgh and Quarterly Reviews: With Addresses and Other Pieces*. London: Longman, Brown, Green, Longmans, & Roberts.

Higgitt, Rebekah. 2007. *Recreating Newton: Newtonian Biography and the Making of Nineteenth-Century History of Science*. London: Pickering & Chatto.

Huxley, Thomas Henry. 1854. "The Vestiges of Creation." *British and Foreign Medico-Chirurgical Review* 13: 425–39.

Huxley, Thomas. 1871. "Presidential Address." In *Report of the Fortieth Meeting of the British Association for the Advancement of Science; Held at Liverpool in September 1870*, lxxiii–lxxxix. London: John Murray.

Lyell, Katherine M. 1881. *Life, Letters and Journals of Sir Charles Lyell*. 2 vols. London: John Murray.

Maxwell, James Clerk. 1879. "Guthrie's Physics." *Nature* 19: 311–12.

Mill, John Stuart. (1831) 1986. "Herschel's Preliminary Discourse." In *Collected Works of John Stuart Mill*, vol. 22, *Newspaper Writings, December 1822–July 1831*, edited by Ann P. Robson and John M. Robson, 284–87. Toronto: University of Toronto Press.

Mill, John Stuart. 1835. "Professor Sedgwick's Discourse; State of Philosophy in England." *London Review* 1: 94–135.

Morrell, Jack, and Arnold Thackray. 1981. *Gentlemen of Science: Early Years of the British Association for the Advancement of Science*. Oxford: Clarendon Press.

O'Connor, Ralph. 2007a. *The Earth on Show: Fossils and the Poetics of Popular Science, 1802–1856*. Chicago: University of Chicago Press

O'Connor, Ralph. 2007b. "Young-Earth Creationists in Early Nineteenth-Century Britain? Towards a Reassessment of 'Scriptural Geology.'" *History of Science* 45: 357–403.

Owen, Richard. 1860. "Darwin on the Origin of Species." *Edinburgh Review* 111: 487–532.

Paley, William. (1794) 1838. *Evidences of Christianity*. Vol. 2 in *The Works of William Paley*. London: Longman and Co.

Paley, William. 1838. *Sermons, Doctrinal, Moral, and Miscellaneous, Continued*. Vol. 4 in *The Works of William Paley*. London: Longman and Co.

Peacock, George. 1841. *Observations on the Statues of the University of Cambridge*. Cambridge: J. and J. J. Deighton.

Peacock, George. 1845. "Presidential Address." In *Report of the Fourteenth Meeting of the British Association for the Advancement of Science; Held at York in September 1844*, xxxi–xlivi. London: John Murray.

Reidy, Michael S. 2008. *Tides of History: Ocean Science and Her Majesty's Navy.* Chicago: University of Chicago Press.

Richards, Evelleen. 2017. *Darwin and the Making of Sexual Selection.* Chicago: University of Chicago Press.

Rudwick, Martin J. S. 2005. *Bursting the Limits of Time: The Reconstruction of Geohistory in the Age of Revolution.* Chicago: University of Chicago Press.

Rupke, Nicolass A. 1994. *Richard Owen: Victorian Naturalist.* New Haven: Yale University Press.

Schneewind, J. B. 1977. *Sidgwick's Ethics and Victorian Moral Philosophy.* Oxford: Oxford University Press.

Secord, James A. 2003. *Victorian Sensation: The Extraordinary Publication, Reception, and Secret Authorship of Vestiges of the Natural History of Creation.* Chicago: University of Chicago Press.

Secord, James A. 2014. *Visions of Science: Books and Readers at the Dawn of the Victorian Age.* Chicago: University of Chicago Press.

Sedgwick, Adam. (1830) 1826–32. [Presidential Address, 1830]. *Proceedings of the Geological Society of London*, 3rd ser., 1: 187–212.

Sedgwick, Adam. (1831) 1826–32. "Address to the Geological Society, Delivered on the Evening of the 18th of February 1831, by the Rev. Professor Sedgwick, M.A. F.R.S. &c. on Retiring from the President's Chair." *Proceedings of the Geological Society of London*, 3rd ser., 1: 281–316.

Sedgwick, Adam. 1833. *A Discourse on the Studies of the University.* London: John W. Parker.

Sedgwick, Adam. 1845. "Natural History of Creation." *Edinburgh Review* 82: 1–85.

Sedgwick, Adam. 1850. *A Discourse on the Studies of the University.* 5th ed. Cambridge: John Deighton.

Sidgwick, Henry. 1876. "Philosophy at Cambridge." *Mind* 1: 238–42.

Snyder, Laura J. 2006. *Reforming Philosophy: A Victorian Debate on Science and Society.* Chicago: University of Chicago Press.

Snyder, Laura J. 2011. *The Philosophical Breakfast Club: Four Remarkable Friends Who Transformed Science and Changed the World.* New York: Broadway Books.

Stair Douglas, Janet Mary. 1881. *The Life and Selections from the Correspondence of William Whewell.* London: C. Kegan Paul & Co.

Tabb, Kathryn. 2016. "Darwin at Orchis Bank: Selection After the *Origin.*" *Studies in History and Philosophy of Biological and Biomedical Sciences* 55: 11–20.

Todhunter, Isaac. 1876. *William Whewell, D.D., Master of Trinity College Cambridge: An Account of His Writings with Selections from His Literary and Scientific Correspondence.* 2 vols. London: Macmillan and Co.

Van Dongen, Jeroen, and Herman Paul, eds. 2017. *Epistemic Virtues in the Sciences and the Humanities.* Boston Studies in the Philosophy and History of Science 321. Cham, Switzerland: Springer.

Walford, Edward, ed. 1863–68. "Charles Darwin." In *Portraits of Men of Eminence in Literature, Science, and Art; with Biographical Memoirs*, 5:49–52. London: Alfred William Bennett.

Whewell, William. 1828. *An Essay on Mineralogical Classification and Nomenclature.* Cambridge: Cambridge University Press.

Whewell, William. 1831. "Lyell's *Principles of Geology.*" *British Critic* 9: 180–206.

Whewell, William. 1832. "Lyell's *Geology,* Vol. 2—Changes in the Organic World Now in Progress." *Quarterly Review* 47: 103–32.

Whewell, William. 1834. "Address." In *Report of the Third Meeting of the British Association for the Advancement of Science; Held at Cambridge in 1833,* xi–xxvi. London: John Murray.

Whewell, William. 1837. *History of the Inductive Sciences, from the Earliest to the Present Times.* 3 vols. London: John W. Parker.

Whewell, William. 1838. *On the Foundation of Morals: Four Sermons Preached Before the University of Cambridge, November 1837.* London: John W. Parker.

Whewell, William. 1840. *The Philosophy of the Inductive Sciences.* 2 vols. London: John Parker.

Whewell, William. 1845. *The Elements of Morality: Including Polity.* 2 vols. London: John W. Parker.

Whewell, William. 1846. *Indications of the Creator.* 2nd ed. London: John Parker.

Whewell, William. 1847. *History of the Inductive Sciences, from the Earliest to the Present Times.* 3 vols. New ed. London: John W. Parker.

Wilberforce, Samuel. 1860. "Darwin's *Origin of Species.*" *Quarterly Review* 108: 225–64.

Yeo, Richard. 1993. *Defining Science: William Whewell, Natural Knowledge and Public Debate in Early Victorian Britain.* Cambridge: Cambridge University Press.

2

Seeing Science as a Communal Practice

MacIntyre, Virtue Ethics, and the Study of Science*

Thomas A. Stapleford and Daniel J. Hicks

Introduction: Science as a Vocation

Speaking before the Physical Society of Berlin in 1918 as part of its celebration of Max Planck's sixtieth birthday, Albert Einstein began his address with a striking image:

> In the temple of science are many mansions, and various indeed are they that dwell therein and the motives that have led them thither. Many take to science out of a joyful sense of superior intellectual power; science is their own special sport to which they look for vivid experience and the satisfaction of ambition; many others are to be found in the temple who have offered the products of their brains on this altar for purely utilitarian purposes. Were an angel of the Lord to come and drive all the people belonging to these two categories out of the temple, the assemblage would be seriously depleted, but there would still be some men, of both present and past times, left inside. (Einstein 1954, 224)

It was true, Einstein conceded, that separating these scientific sheep from the goats would be a "ticklish job," and indeed, some of those expelled would have been "largely, perhaps chiefly, responsible for the buildings of the temple of science" (224). Nonetheless, he also insisted that the temple—as a temple—would not have existed without those left inside, for they were the ones devoted to science itself, and not just to the benefits that success in science might bring; therefore, it was they who gave science its particular character. For those driven out, Einstein argued, "any sphere of human activity [would] do, if it comes to a point; whether they become engineers, officers, tradesmen, or scientists depends on circumstances" (225).

That same year, the German sociologist Max Weber depicted a similar contrast in a lecture to a group of undergraduates at the University of Munich. Titled "Wissenschaft als Beruf," Weber's lecture has been traditionally translated as "Science as a Vocation," and rightly so, for Weber took as his task an examination

Thomas A. Stapleford and Daniel J. Hicks, *Seeing Science as a Communal Practice* In: *Science, Technology, and Virtues*. Edited by: Emanuele Ratti and Thomas A. Stapleford, Oxford University Press. © Oxford University Press 2021. DOI: 10.1093/oso/9780190081713.003.0003 of what it meant to pursue "'science for science's sake,' and not merely because others, by exploiting science, bring about commercial or technical success" (Weber 1946, 138). The latter, Weber declared, were equivalent to "greengrocers" who gathered and traded knowledge in order to gain something else. To pursue science as a vocation meant to value science for its own direct benefits—what Weber characterized as a kind of intellectual "clarity." Through coming to understand both humans and the nonhuman world, a scientist could press someone "to give himself an *account of the ultimate meaning of his own conduct*," a task that stood "in the service of 'moral forces'" (Weber 1946, 149, 151–52, emphasis in original).

Writing about American physicists in the 1950s and 1960s, David Kaiser documented a parallel phenomenon, with an older generation of physicists lamenting that their new students were pursuing physics just for financial stability and security. Indeed, compared to their peers entering other fields, prospective physicists were more likely to list "making a lot of money" as a top career objective and were far more skeptical of what Weber had called "science for science's sake." As one graduate student put it, "I don't know about all of this romanticizing of physics. As far as I'm concerned, physics is no different from any other occupation. It is work. Like every other kind of work, it's a job. I'm interested in it because it so happens you can get very good jobs in physics that pay a lot of money" (Kaiser 2004, 866–71, quote on 870).

Romanticized or not, the view of science as more than just a job—more than just a means of making money—has persisted, even in the face of major concerns about the excessive commercialization of science. For example, a 2014 survey of members of the American Association for the Advancement of Science found that only 4 percent of respondents cited "practical issues" such as funding or job opportunities as among the "most significant" factors in their career choice; instead, the most common items were "intellectual challenge, lifelong curiosity, and love of science or nature" (Pew Research Center 2015, 69). A close study of eighteen government research scientists in the United Kingdom found similar results, with the most common motivations for research being "curiosity" or "doing good science"; only one person mentioned money (Jindal-Snape and Snape 2006, 1332). Perhaps remarkably, even among academic scientists who have engaged in commercial research, less than one-third reported boosting their personal income as an "important" or "very important" motivating factor (Lam 2011, 1361). Though we might doubt the accuracy of scientists' reports about their motivations, at a minimum these results illustrate the persistence of a vocational ideal for science, of a conviction that science *ought* to be about "intellectual challenge, lifelong curiosity, and a love of science or nature."

Taken together, these vignettes share several common features. First, they document a recognized difference between what we might loosely call science-as-an-end ("science for science's sake," in Weber's terms) and science-as-a-means (a way to secure money, prestige, power, or something beyond science itself). Second, they reveal divisions within scientific communities with respect to those motivations: though some practitioners promote a vision of science-as-an-end, not all would agree. Einstein suggests that a winnowing by the angel of the Lord would leave the temple of science "seriously depleted"; the physicists Kaiser studied complain about their younger colleagues; and although the 2014 Pew survey found "intellectual challenge" et cetera as the most dominant motivational category, it was still cited by only about a third of the respondents (Pew Research Center 2015, 69). Finally, the earlier three vignettes give this distinction a highly moral tone. Einstein used explicit religious imagery; Weber contrasted the "service of 'moral forces'" of vocational science with the bourgeois "green-grocer"; and the physicists in Kaiser's study saw their money-seeking acolytes as deficient in character. Such characterizations match what we would claim are more general features of modern science: that many scientists recognize a distinction between pursuing science-as-an-end and pursuing science-as-a-means; that their communities (and perhaps they themselves) may be torn between those two goals; and that these differing goals have been tied to different moral values or ethical choices. We are thus in agreement with Steven Shapin's assertion that the moral dimensions to contemporary science remain as strong as ever, despite common declarations to the contrary (Shapin 2009).

Drawing upon the work of the moral philosopher Alasdair MacIntyre (1984), in a previous essay we identified science-as-an-end as a central component of what we called a "communal practice" (Hicks and Stapleford 2016). Situating science-as-an-end within the larger conceptual framework of communal practices allows us not just to understand why it carries moral overtones but also to clarify how we should interpret that phrase (including in ways that depart from the characterizations of Einstein and Weber). In this chapter, we summarize the concept of a communal practice and explore some of its ramifications. We argue that seeing science as a communal practice allows us to make sense of scientists' own accounts of their motivations and goals without falling into the trap of either idealizing science or dismissing those ideals. Further, we claim that viewing science as a communal practice has important implications for debates about the ethics of science, and notably for the place of virtues in science. In turn, that means that treating science as a communal practice has consequences for how we would wish to train scientists and form scientific communities.

Defining a Communal Practice

To call a set of actions a *practice*, as we use the term here, means to render those actions intelligible by placing them within a narrative that contains an implicit or explicit goal.[1] These narratives can be quite small ("She is brushing away a fly") or large, encompassing many actions ("She is training to become a cellist"). Practices can thus come on a range of scales and can be related in complicated ways.

To call a group of actions a practice is to interpret them, and of course there can be multiple interpretations of the same set of actions. In some cases, those interpretations may be compatible ("I am writing an essay"; "I am making a philosophical argument that I care about deeply"; "I am trying to get tenure"). But in others, they may be opposed ("He is telling the truth"; "He is lying"). Critically, even when interpretations are compatible with one another right now, they may imply different pathways for future actions because they have distinct goals. Right now, for example, "I am making a philosophical argument that I care about deeply" may be compatible with "I am trying to get tenure," but as every early-career scholar knows, that may not always be the case; perhaps what I care about deeply is too ambitious or too uncertain for the time I have remaining before a tenure decision is made.

Communal practices, in our terminology, are a subset of practices that share central features of what Alasdair MacIntyre identified simply as a "practice."[2] Specifically, they are sustained, complex, and collaborative practices in which participants are accountable to one another and whose goals can be characterized as ends-in-themselves. Though scientific research is our focus in this chapter, many other activities can also be understood as communal practices, such as other forms of scholarship, social movements, medical care, raising and educating children, and art. To illustrate the features of a communal practice, we will take several of its core elements in turn.

Complex and Collaborative

By *complex* we mean that a communal practice is built from, and entangled with, a range of other practices. A molecular biologist, for example, may be designing experiments, manipulating equipment, mentoring grad students, analyzing data, writing papers, developing grant proposals, and so forth—all of which may be understood as part of the communal practice of molecular biology. Furthermore, most communal practices involve institutions that enforce norms (which are justified in terms of the goals of the practice) and that provision resources.

Communal practices are *collaborative* rather than individual. The collaborations among practitioners can be tight, as in contemporary research labs, or more diffuse, as in the early modern Republic of Letters. Because communal practices are social by definition, they will also inevitably have a political life (Warren 1999). There may be internal disputes, such as how to provision scarce resources, how goals should be characterized, the acceptability of novel methods or techniques, how students should be trained, and so on. But practitioners will also need to demarcate the boundaries of their community, and they may attempt to use power to promote the interests of their practice over those of other practices or institutions (Gieryn 1983).

Social Accountability

Like all practices, communal practices are goal-oriented, but because they are also collaborative, they have a normative structure at both the individual and group levels. This normativity may be partially embodied by explicit rules, such as the guidelines for conducting an ethically and epistemologically rigorous series of clinical trials in pharmaceutical research. However, there are deep philosophical reasons to think that norms also take other forms, such as concrete models of excellent (or poor) behavior (e.g., Wittgenstein 1963, sec. 185ff; Collins and Evans 2007, ch. 1).

Norms can be introduced and enforced from "outside" of the practice, just as the US government regulates clinical research. However, norms can also develop and be enforced among practitioners themselves. Though these norms are social, practitioners need not share the same norms or the same interpretations of those norms in order to remain part of the same practice. Instead, the crucial feature is the operation of accountability and its politics. As Joseph Rouse argues, practices are characterized fundamentally not by shared rules but by accountability among practitioners (Rouse 2007). Different lab groups may disagree substantially about what constitutes good research, but they are held together as a communal practice by being subject to evaluation by other practitioners. Should the differences grow too great, the practice might fragment, with subgroups limiting the extent to which they are accountable to their former colleagues. Thus, the boundaries of communal practices are established *sociologically and politically* (who is accountable to whom, and for what actions?) rather than by an underlying *philosophical unity* (do these individuals share precisely the same norms and the same interpretation of them?). Obviously, practitioners will compare norms when deliberating about the boundaries of accountability, but ultimately those decisions about accountability will define the boundaries of the practice. Accordingly, the norms in a practice may be like the "family resemblances" described by Wittgenstein: each practitioner's norms might share some commonalities with others or have close resemblances to them, but no single group of norms demarcates the precise bounds of the practice.

Because communal practices are maintained sociologically and politically in this way, we should expect repeated and continual conflict over the norms; in the case of science, those conflicts may involve arguments over what constitutes good knowledge, what techniques or methods are appropriate for producing that knowledge, and what characteristics are needed to be a good scientist. Moreover, we should expect the contours of these arguments—the themes, core issues, areas of tacit consensus, or key points of dispute—to shift over time.

Ends-in-Themselves

The contrast between science-as-an-end and science-as-a-means, so central to the vignettes that opened this chapter, has an analogue in MacIntyre's distinction between *goods of excellence* (or *internal goods*) versus *goods of effectiveness* (or *external goods*). In *After Virtue*, internal goods are defined as those that can only be achieved through proper engagement with a particular communal practice; I can develop a new model for protein folding only by participating in the practice of biochemistry. By contrast, external goods such as power, money, or status can be obtained through a variety of means (MacIntyre 1984, 187ff.). In MacIntyre's later terminology, internal goods are "goods of excellence," whereas external goods are "goods of effectiveness"—they are purely instrumental, means for reaching an end (MacIntyre 1988, ch. 3). MacIntyre recognizes that both types of goods are, indeed, good things to have. Money is necessary to pay for research equipment, status is necessary to recruit lab assistants, and so on. But there are ethically significant, qualitative differences between them.

Taking MacIntyre's discussions of these two types of goods as a whole, goods of excellence can be distinguished by four features: they are progressive, collective, intrinsically valuable, and integrated.[3] Goods of excellence are *progressive* in two senses. First, practitioners are continually striving to reach the ideals of their practice, to produce goods (objects, knowledge, performances, etc.) that are equal to or better than what their community has done in the past. Second, practitioners may debate about the proper norms or goals of their practice, and such debates always take place under the rubric of progress. New norms must in some sense improve over the old; otherwise, why change? In science, we can map these two forms of progress onto Kuhn's (1996) distinction between normal science (making progress within a paradigm of standards, core questions, and values) and revolutionary science (making progress across paradigms). Though Kuhn insisted that the notion of progress across paradigms has no objective meaning because different paradigms posit competing notions of progress, he nonetheless acknowledged that scientists must subjectively view a new, favored paradigm as superior to the old (160–73).

Goods of excellence are *collective* because (unlike money, power, or status) my possession of them does not inhibit you from possessing them to the same degree; moreover, by obtaining these goods I enhance the scope or development of the practice as a whole. The excellence of synthesizing a new compound does not prevent someone else from synthesizing that same compound, and if I share my knowledge with the community, I have advanced the practice as a whole. Of course, I could choose to keep my knowledge secret so as reap greater monetary rewards—but in that case, I am treating the synthesis as a "good of effectiveness," as a means to other ends.

If goods of effectiveness are characterized as means to an end, then goods of excellence must be ends in themselves; that is, they must have *intrinsic value*, at least according to the norms of the practice. That means that the possession of a good of excellence must be valuable on its own, independently of its value for achieving other goods (Zimmerman and Bradley 2019). The notion of intrinsic value is apt to be misunderstood, so a few clarifications are in order.

First, intrinsic value per se does not define the boundary between "basic" and "applied" science, much less between science and technology. Aerospace engineers designing sleek, functional, and efficient aircraft may argue that excellent design and construction have intrinsic value; an elegant plane has an aesthetic value akin to a fine work of art. Instead, when comparing goods of excellence to goods of effectiveness, the core distinction is between goods that are constitutive of human flourishing (what Aristotle called *eudaimonia*) and those that are merely instrumental—that is, those that may enable flourishing here and now but are not constitutive parts of such flourishing. For example, those who value the production of beautiful and elegant artifacts may see such production as constitutive of human flourishing; the production of art thus has intrinsic value. They may therefore patronize the arts, working in their own field to produce money as an instrumental good (a good of effectiveness) that may sustain an artist and allow for the further development of goods of excellence.

Naturally, goods with intrinsic value may also have instrumental value. As we noted earlier, placing a set of actions within a given practice is an interpretation, and most actions can support multiple interpretations. Suppose I am conducting research on the role of cellular enzymes in cancer metastasis (e.g., Hawk et al. 2018). I may have a genuine desire to understand cellular physiology that is independent of what else that knowledge might enable me to accomplish (knowledge of physiology is the good of excellence). Simultaneously, I may wish to help reduce cancer mortality, in which case such physiological knowledge is just an instrumental component within a broader communal practice of medical research (knowledge of how to sustain human health is the good of excellence). And I may also be driven in part by the desire for recognition or financial success (goods of effectiveness).

Though these three narratives may all fit my actions at this time, they may also pull me in different directions. Equally important, they may pull a community in different directions, leading to political struggles over the goals of the practice, and indeed over its status as a communal practice (and not just a path for achieving goods of effectiveness). The conflict over the future of postwar American physics that David Kaiser (2004) described (wherein the older generation complained that their younger counterparts sought only financial security) represents precisely this kind of struggle.

Goods of excellence can have intrinsic value because they are integrated into broader conceptions of what constitutes a meaningful life, a form of human excellence, or a worthy mode of being-in-the-world. To promote science-as-an-end makes sense only if one sees the development and possession of scientific knowledge as one component of an ethos that, tacitly or explicitly, characterizes a desirable form of life. Otherwise, the goods of science would only be instrumental, means toward an excellent life rather than a constitutive part of such a life. Of course, there may be other goods of excellence required for the "good life," and at times these other goods may come into conflict with the practice of science. But any talk of intrinsic value must entail broader ethical claims. That is precisely why, of course, Einstein laced his mythical account of science with religious imagery and why Weber described science as a vocation in such deeply moral terms.

Virtues

MacIntyre introduces his theory of communal practices to reveal the nature and function of virtues. Once we understand the normative character of communal practices, we can recognize that participants must have certain qualities of character in order to realize a practice's goods of excellence, and that these qualities are themselves further goods of excellence of the practice. Drawing on Aristotle, MacIntyre theorizes these qualities as virtues: dispositions to think, act, and feel in ways that exemplify the models of excellence in the practice and allow it to be sustained and advanced (MacIntyre 1988, 12–145).

Since discussions of virtue by ethicists often employ the term differently than contemporary colloquial usage, a few clarifications are in order. First, virtues in the Aristotelian sense are not limited to what we might call "moral" virtues. The Greek term translated as "virtue," *arete*, means "excellence." The virtuous practitioner, in this respect, is the excellent practitioner, the one whose qualities allow her to routinely realize the goods of the practice. Aristotelian virtues are thus the qualities of excellence in a practice; they may refer to the physical or intellectual capabilities of a practitioner, how the practitioner relates to her environment (including other practitioners), how she assesses the concrete situation in which she finds herself, or how she reacts to that assessment. Many of these qualities may have no bearing on what we commonly identify as "moral": the virtuous (excellent) long-distance runner, for example, will be able to run longer distances with less physical strain than an untrained counterpart, a trait that is surely admirable but rarely considered "moral."

Second, virtues may build upon existing tendencies, but they are nonetheless acquired traits. Becoming virtuous requires building dispositions or habits to act in certain ways, which requires emulation, repetition, and correction. Virtue is learned through practice in at least two senses. First, virtue is learned by *practicing being virtuous*, rather than through a primarily intellectual process of reflecting on moral duties or calculating aggregate utilities. For many virtue ethicists, virtues are cultivated before they are understood; we learn habits of honesty starting as toddlers, and only later do we learn why honesty is a valuable habit (Hursthouse 1999, ch. 5). Second, virtue is learned *in and through our relationships with others* in our communal practices. Paradigmatically, as novices in a communal practice we emulate our teachers and mentors, and receive feedback from them that helps correct our behavior. Consequently, the social relations and institutions that are part of a communal practice must facilitate the emulation, repetition, and correction of its virtues. Historians and sociologists of science have shown that scientific training involves exactly this process of emulation, repetition, and correction (Buchwald 1995; Warwick 2003; Kaiser 2005).

Third, unlike some contemporary uses of the concept of "habit," virtues are not *routines*, actions or series of actions that are performed automatically. Julia Annas makes this point by likening virtues to skills such as playing the piano (Annas 2011, chs. 2–3). A skilled pianist has cultivated certain habits through many hours of emulation and repetition, such as fingerings for standard scales and arpeggios. The pianist can play these with little conscious thought or attention. But that mindlessness enables her to direct her attention to other, more artistically significant aspects of her playing: "better ways of dealing with transitions between loud and soft, more subtle interpretations of the music, and so on." This conscious artistry, in turn, shapes the way she plays the scales and arpeggios, so that even these habits are not exercised as mere routines. In the same way, virtues are habits that "enable us to respond in creative and imaginative ways to new challenges" (14–15).

Finally, in a departure from MacIntyre, in this chapter we adopt a weaker approach in which virtues are always relative to a practice. Thus what is a virtue in one practice may be indifferent or even vicious in another. For example, some feminist standpoint theorists argue that a degree of sympathetic identification with one's research subjects is essential for good social science, perhaps even in fields such as corn genetics (Keller 1983; Ladner 2002), but this sympathy conflicts sharply with prevailing notions of objectivity as detached and disinterested (Jordan, Gust, and Scheman 2011). A theory of communal practices, therefore, does not provide us with a full ethical theory. It distinguishes between goods of excellence and goods of effectiveness, but it does not on its own terms help us to differentiate between various goods of excellence that might be proposed.

Communal Practices and the Analysis of Science

In previous work (Stapleford and Hicks 2016), we showed how viewing science as a communal practice illuminated certain historiographic trends and debates while also highlighting the ethical commitments of a historian and clarifying the relationship between the history of science and science itself. Here we aim to explore the more general value of using the concept of communal practices to examine science.

Taking Myths (and Scientists) Seriously

We saw in the introductory section that scientists continue to list "curiosity" and "doing good science" as primary motivations for their own work, viewpoints that resonate with what Weber called "science for science's sake" and with Einstein's remnant in the temple. Yet in a more cynical age, and in a time where scientific research often requires massive amounts of funding and thus justifications to funders (whether university administrators, companies, or legislators), it becomes more difficult to see "science for science's sake" as anything other than a "romanticized" delusion, to quote one of the postwar American physicists cited by Kaiser (2004, 870). Indeed, writing in 2005, the biologist Adam Liska dismissed claims about the intrinsic value of scientific knowledge as a "myth," insisting that modern science was overwhelmingly instrumental and that the only question was about the ends toward which science would be directed—for example, self-aggrandizement, military power, commercial exploitation, or "the curing of disease and the advancement of human life through technology" (Liska 2005, 12).

While recognizing the structural changes that have affected science funding and institutional context over the last hundred years (Mirowski and Sent 2002), historians have likewise complicated any view of earlier eras as bastions of selfless, rarified intellectual pursuit. For example, the ties that would develop between nineteenth-century physical sciences and industrialization (Haber 1958; Smith 1998; Lucier 2010) were anticipated and celebrated already in the seventeenth century by Francis Bacon. Likewise, Galileo's pursuit of patronage and his bitter priority disputes (Biagioli 1993) seem much more the norm for science than the exception. Science was "never pure," as Steven Shapin (2010) aptly put it.

Nevertheless, it is one thing to recognize the complexity of individual motivations and the messy tensions that necessarily accompany the pursuit of any practice; it is another to deny the legitimacy of goods of excellence altogether. A communal practice can never be "pure" if by purity we mean excluding the pursuit of goods of effectiveness altogether. Quite to the contrary, MacIntyre insisted that goods of effectiveness are necessary for the flourishing of even a communal practice because they are the resources that make the practice possible. (That is why they are goods of *effectiveness.*) Likewise, scientific knowledge is not the only good of excellence, and such knowledge may legitimately be instrumental for other ends. But none of this reasoning implies that we can simply erase the distinction between goods of excellence and goods of effectiveness, or that we can productively impose an analytic framework that deliberately regards all science as *merely* instrumental.

Regrettably, some scholars in science and technology studies have done just that. For example, the eminent French sociologist Pierre Bourdieu insisted that "all practices, including those purporting to be disinterested or gratuitous, and hence non-economic, [can be treated] as economic practices directed towards the maximizing of material or symbolic profit"—that is, toward goods of effectiveness (Bourdieu 1977, 183). Bruno Latour and Steve Woolgar's pioneering ethnographic study, *Laboratory Life*, one of the classics of early science and technology studies, argued that one could ignore scientists' professed motivations and treat all of their activities as a never-ending quest for "credibility," their version of Bourdieu's "symbolic capital" (Latour and Woolgar 1979, 198, 207–8). Not surprisingly, this outlook has taken particular hold within efforts to use neoclassical economic theory to analyze science. For example, Fernández Pinto (2016) argues that recent applications of economic methods to model the social dimensions of scientific inquiry by using rational choice or game theory all assume that scientists' "nonepistemic motives" can be nothing more than the self-interested pursuit of "personal credit and recognition" (464). Essentially, such approaches treat scientists' statements about their own passion for knowledge as either a rhetorical smokescreen or a self-delusion, something that can be ignored without loss of understanding.

Fortunately, the concept of a communal practice—and practice theory more generally—offers us a way to take scientists' accounts seriously (indeed, to understand their importance) without ignoring the complexity of motivations. Once we understand a practice as an interpretation, a narrative that can both guide future actions and render past actions intelligible, we can see that both Bourdieu's notion of "practice" and scientists' own claims about science as a vocation are stories, narratives about the meaning and goals of science. We can recognize that multiple stories might be consistent with certain actions, and we can then begin to interrogate them on a different level. We can ask what these stories reveal (and conceal), and, most importantly, what function they serve—why they are being told and how taking a given narrative seriously might shape our future actions. What would be the implications for practitioners of taking Bourdieu's account seriously? Or Weber's? To characterize science as a communal practice is not to insist that goods of excellence are the only goals that motivate all scientists, or even a single scientist; it is to insist that goods of excellence should be at the heart of the practice, that to lose sight of them is to undermine the integrity of the practice itself.

Liska (2005) was right to call "science for science's sake" a myth, but not in the sense of it being false. Rather, it is a story that attempts to capture an important aspect of reality in order to reorient us and guide our actions. It was fitting, therefore, that Einstein's address on this topic was explicitly mythological, from its timeless temple to its "angel of the Lord," for Einstein was trying to lay out one potential narrative about science (not the only one!) and its ramifications. To tell the story of any practice as a communal practice is to place it in a particular ethical framework, which is both why scientists tell such stories and why we analysts need to take them seriously.

Communal Practices and the Purity of Science

If the framework of communal practices helps us to understand the significance of these mythical narratives, it also highlights the limits of existing debates about the "purity" of science. Both advocates of purity (such as Einstein and Weber) and critics (such as Liska) often create a false dichotomy between epistemic and practical ends, as though these were mutually exclusive. Furthermore, some advocates (including again Einstein and Weber) and some critics (such as Bourdieu) fail to distinguish between various practical ends, treating all of them as if they were (in MacIntyre's terms) goods of effectiveness. Thus Einstein casts out those who conduct research for "utilitarian purposes" and Weber treats the pursuit of "technical success" as equivalent to being a "greengrocer." Given these two assumptions, either science is a communal practice that pursues knowledge as a good of excellence, or science pursues only goods of effectiveness and so is not a communal practice.

Neither of these assumptions is necessary. Again, a single good can be pursued by the same individual for multiple reasons. In fields such as agriculture, biomedicine, conservation biology, and environmental science, many academic researchers are motivated by both intellectual curiosity and the desire to improve the world—to grow more and more nutritious food, to treat disease, to protect endangered species, to remediate pollution. The knowledge they produce aims to be both epistemically valuable and practically useful.

Nonetheless, these twinned epistemic and practical aims do create the possibility for dilemmas. In areas such as pharmacology and development economics, concerns about rigor—an epistemic virtue—have led to a heavy emphasis on randomized controlled trials (RCTs). A properly conducted RCT can have high *internal validity* or *reliability*, giving us high confidence in claims about causal effects *within the study sample*. But they do not ensure *external validity* or *relevance*; that is, a RCT cannot tell us whether its results can be extrapolated to the population we actually want to treat (e.g., Cartwright 2007). How should scientists trade off the epistemic ends of rigor or internal validity against the practical ends of producing a relevant treatment? Similarly, Daniel Steel considers "Ibsen predicaments," in which the practical ends of research seem to require sacrificing its epistemic ends (Steel 2017). Recognizing these tensions as fundamental ethical dilemmas—conflicts between two intrinsic goods—can be descriptively or analytically useful. How have particular communities of scientists navigated such dilemmas? Specifically, how have ideas such as the distinction between "pure" and "applied" or a social contract for science been deployed by scientists in the politics surrounding these dilemmas?

The second assumption often shared by both defenders and critics of "purity" is the idea that the practical ends of science are not intrinsically valuable, and hence that the critical ethical line is between knowledge-for-understanding and knowledge-for-action. Whether the point is to defend or critique the pursuit of pure knowledge, the contrasting practical ends are money, status, power, and other goods of effectiveness. But consider again fields such as agriculture, biomedicine, conservation biology, and environmental science. Though these are very much applied fields focused on "technical success," their constitutive ends are central components of a good human life or *eudaimonia* (Hicks, Stahmer, and Smith 2018), just as we argued earlier that the production of beautiful art can be constitutive of human flourishing. The goods pursued by these applied fields are thus intrinsically valuable and not merely generic means to further ends.

Therefore, even if the knowledge produced by scientific research were regarded as merely instrumentally valuable, such research could still have intrinsic goods as its constitutive ends, and so could still be a communal practice. Normatively, the significant contrast is not between pure and applied scientific research, but instead between research governed by goods of excellence and research governed by goods of effectiveness. This is the ethical difference between goods such as health and a pollution-free environment, on the one hand, and goods such as money and status, on the other.

Of course, it is one thing to say that a body of scientific research is engaged in the pursuit of goods of excellence, and another thing for it to actually do so. This is the difference between the nominal and effective ends of a practice. Arguably, fields such as pharmacology and agronomy are caught in a mismatch between their nominal and effective ends. Nominally, the aim of pharmacology is to produce new drugs to treat disease, and the aim of agronomy is to feed the world. But often it appears that research is not conducted in ways that actually promote these aims. Globally, the distribution of biomedical research is almost completely unrelated to the distribution of disease burden, with disproportionate attention to cancer and neglect of respiratory diseases and nutritional deficiencies (Evans et al. 2014). Moreover, the design, conduct, and publication of clinical trials often seem to have as their primary goal the gaming of the regulatory system rather than the production of reliable and relevant knowledge (Holman and Elliott 2018).

In these cases, the concerns of Einstein and Weber seem apt: these areas of research are not actually pursuing any goods of excellence but instead are subordinating intrinsic goods (such as knowledge, human health, and human nutrition) to goods of effectiveness (namely, corporate profit). They are, in short, not communal practices.

A MacIntyrean perspective on communal practices emphasizes that goods of effectiveness are necessary for the continued existence of a communal practice. We cannot do scientific research on any meaningful scale without money, status hierarchies, or internal and external political struggles. But normatively we also cannot allow goods of effectiveness to dominate and displace the goods of excellence that provide the primary reason for pursuing scientific research. When goods of effectiveness do come to dominate science, scientific endeavor loses not so much its "purity" as its ethical integrity.

Sustaining Science as a Communal Practice

We have argued that seeing science as a communal practice has ethical ramifications, and indeed that it is essential for maintaining the integrity of the practice. Anyone who supports those goals, therefore, has an interest in maintaining a vision of science as a communal practice (even though that vision might coexist among others) and supporting that outlook among its practitioners. But how can that be accomplished?

Most directly, of course, scientists committed to the communal practice can continue to articulate that vision, much as Einstein or Weber did (though we would advocate a more inclusive scope that could encompass so-called applied sciences). But speeches only go so far. What we need are practitioners who are committed to treating science as a communal practice, who will strive for that ideal, and whose actions will be guided by that vision. In short, to sustain science as a communal practice, we need virtuous practitioners (i.e., practitioners with qualities defined as virtuous from within that practice).

Of course, whether and how virtue can be cultivated (much less effectively taught) has been a core philosophical debate since antiquity. Over the last fifteen years, that debate has been greatly enlivened and enriched by dialogues between philosophers and empirical psychologists (e.g., Russell 2009; Snow 2010, 2014; Annas, Narvaez, and Snow 2016; Masala and Webber 2016; Kristjánsson 2018). Though this literature is too broad and diverse to summarize fully here, one common and promising approach has been to see the cultivation of moral virtues as psychologically akin to the cultivation of skills (e.g., Narvaez and Lapsley 2005; Russell 2009; Snow 2010; Annas 2011). From an Aristotelian point of view, that makes sense since moral virtues are a subset of a broader set of virtues (excellences), including intellectual capabilities, craft skills, and so forth (Russell 2014).[4] Recognizing these ties, though, means that much of the psychological literature on the development of expertise can be relevant to our understanding of moral virtue as well.

Several themes emerge from these reflections. For example, whereas skills can sometimes be partially developed by relying on direct feedback from the environment (did that shot go in or not?), they may also require—and moral virtues will always require—emulating an existing expert, and perhaps receiving direct guidance. As Snow (2016) has argued, novice practitioners need not explicitly be seeking to develop moral virtues; they may only wish to become better at their specific role—for example, better teachers, nurses, or woodworkers. But insofar as the expert from whom they are learning views her respective domain as a communal practice, a novice practitioner will also be emulating someone striving to be virtuous and hence seeking to emulate those traits. Nonetheless, by explicitly discussing the qualities one is attempting to foster and why, experts can make the pursuit of virtue a self-conscious activity and more readily build habits of critical self-reflection (Annas 2011). Substantial research suggests that people are more likely to behave in virtuous ways when they are routinely prompted to reflect on their actions within an ethical framework (Athanassoulis 2016, 222–25), a finding that of course resonates with a wide range of traditional religious practices (daily devotional readings, habitual prayer, examinations of conscience, etc.).

Beeler and Bezuidenhout (2018) examine what this might look like in science through a study of the virtue of "docility" (the traditional term for a proper mean between excessive skepticism of another person's claims and excessive trust). Within science, developing the proper virtue of docility is essential not just for evaluating research literature but also, as Beeler and Bezuidenhout emphasize, for proper collaboration and training within a lab. For example, they draw on ethnographic research to show how common practices such as journal clubs can be valuable forums for graduate students and postdocs to learn how to exercise docility appropriately—how to read critically, how to interrogate a paper, and how to use this process to strengthen their own writing and research presentations. But Beeler and Bezuidenhout also argue that the traditional structure of undergraduate laboratory science education—usually focused on uncritical repetition of lab techniques—tends to produce an imbalanced deference to authority. From their own case study, they present examples where informal (and incorrect) protocols were passed down informally through a lab without being questioned until they were (unintentionally) brought to the attention of the lab director (226–30). Beeler and Bezuidenhout suggest that addressing this imbalance requires bringing into laboratory practices the same kind of explicitly probing and critical reflection that characterizes journal clubs, a task that might require rethinking undergraduate laboratory experiences and instructor training (233–35).

Equally important are the many informal ways in which senior scientists can promote a powerful conception of the qualities of an excellent scientist through their behavior and commentary, including stories, jokes, and evaluations of colleagues, postdocs, or students. Furthermore, because senior scientists can have great influence on grad student placement and postdoctoral opportunities, they can likewise reinforce these standards. Traweek's (1992) rich ethnographic study of US particle physics laboratories in the 1980s illustrates this beautifully—provided that we remember that the "virtues" of a particular communal practice may not be what we would wish to regard as moral virtues more broadly. In Traweek's account, the ideal particle physicist was both explicitly and implicitly presented as brash, independent, competitive, "superficially nonconformist," disdainful of perceived intellectual inferiors, intensely dedicated to their research, and possessed of a "childlike egoism" (85–93, esp. 87, 91). Young physicists who failed to cultivate the appropriate persona could be marginalized or culled from the field.

Of course, we may find this "ideal" particle physicist distressing or even morally repugnant. Recall that the concept of a communal practice that we use here makes a weak ethical category: for example, it distinguishes the true Nazi ideologue from the conformist who joins the Party merely to get ahead, but it says nothing about the moral turpitude of racism. Nonetheless, the concept of a

communal practice—with its links to virtue ethics and to practice theory in sociology and philosophy—does provide us with resources to critique and potentially alter a deviant communal practice.

Because the goods of excellence of a communal practice must be integrated
with a broader vision of what constitutes a good life, any disjuncture between the two must be addressed. In the case of Traweek's particle physicists,
for example, the split between the characteristics of the ideal physicist and
more traditional moral virtues suggests either that the traditional moral
virtues are not actually true virtues (and thus should be adjusted to match
the ideal traits of physicists), that particle physics is not a communal practice (and thus the mismatch is untroubling), that the goods of excellence for
US particle physics have been misspecified or misordered (and hence that
correcting them would alter the perceived virtues of the practice), or that US
particle physicists have misjudged the qualities required to sustain their field
as a communal practice. Assuming prima facie that particle physics is a genuine communal practice, the last of these seems a promising place to start.
This is precisely how, for example, feminist critiques of archaeology in the
1980s and 1990s succeeded in altering the practices of their field: by showing
how a feminist perspective could actually enhance the epistemic goods of
excellence predominately male archaeologists were already pursuing (Hicks
2014, 3291).

Having made such a critique, the dialogue between virtue ethics and empirical psychology suggests how to proceed: create procedures that foster new
habits and thus new patterns of action, and cultivate critical self-reflection on
those actions; ensure that mentors routinely talk explicitly about the goals of the
practice and how they are linked to both patterns of actions and virtuous qualities; and create environments where practitioners can freely examine and discuss
ethics and bring those reflections to bear on their practice (Athanassoulis 2016,
220–26).

Conclusion

To see science as a communal practice is to see it as having its own internal
goals, "goods of excellence" that are perceived as having value in themselves
and that are integrated into broader visions of the good life for human beings.
To pursue those goods is to pursue one way of living such a good life, and it
demands that an individual possess certain qualities, the virtues, that are constitutive of such a life.

That outlook resonates with how some scientists have described their own
work both now and in the past—namely, as a vocation and not merely as a job.

In many cases, this vision is compatible with other narratives about science—a tool for technological progress, a space to demonstrate one's intellectual prowess, a way to make money, and so forth. But at times it may also deviate from them in ethically meaningful ways. To recognize this vision, to understand the dynamics that underlie it, that support it, and that may challenge it, is thus crucial to an ethically rich analysis of science itself.

Notes

* Portions of this essay appeared in modified form in Daniel J. Hicks and Thomas A. Stapleford, "The Virtues of Scientific Practice: MacIntyre, Virtue Ethics, and the Historiography of Science," Isis: Journal of the History of Science Society 107, no. 3 (2016): 449–72. Stapleford's work on this project was supported by a grant from the Templeton Religion Trust, "Developing Virtues in the Practice of Science."

1. For an overview of theoretical approaches to "practice," see Nicolini 2013.
2. See MacIntyre 1984, 187ff. For a general introduction to MacIntyre's concept, see Knight 2008. We have opted for the term "communal practice" to distinguish MacIntyre's view from other accounts.
3. These terms are ours rather than MacIntyre's, but they reflect an attempt to synthesize key elements of his arguments.
4. This is not, of course, to elide the differences between skills and moral virtues that Aristotle describes.

References

Annas, Julia. 2011. *Intelligent Virtue*. Oxford: Oxford University Press.

Annas, Julia, Darcia Narvaez, and Nancy Snow, eds. 2016. *Developing the Virtues: Integrating Perspectives*. New York: Oxford University Press.

Athanassoulis, Nafsika. 2016. "The Psychology of Virtue Education." In *From Personality to Virtue: Essays on the Philosophy of Character*, edited by Alberto Masala and Jonathan Webber, 207–28. Oxford: Oxford University Press.

Beeler, Dori, and Louise Bezuidenhout. 2018. "Docility Is Not Passiveness: Teaching Learners to Learn in Science Education." *Philosophy, Theology and the Sciences* 5, no. 2: 216–38.

Biagioli, Mario. 1993. *Galileo, Courtier: The Practice of Science in the Culture of Absolutism*. Chicago: University of Chicago Press.

Bourdieu, Pierre. 1977. *Outline of a Theory of Practice*. Cambridge: Cambridge University Press.

Buchwald, Jed Z. 1995. *Scientific Practice: Theories and Stories of Doing Physics*. Chicago: University of Chicago Press.

Cartwright, Nancy. 2007. "Are RCTs the Gold Standard?" *BioSocieties* 2, no. 1: 11–20.

Collins, Harry M., and Robert Evans. 2007. *Rethinking Expertise*. Chicago: University of Chicago Press.

Einstein, Albert. 1954. "Principles of Research." In *Ideas and Opinions by Albert Einstein*, edited by Carl Seelig, translated by Sonja Bargmann, 224–27. New York: Crown Publishers.

Evans, James A., Jae-Mahn Shim, and John P. A. Ioannidis. 2014. "Attention to Local Health Burden and the Global Disparity of Health Research." *PLOS ONE* 9 (4): e90147.

Fernández Pinto, Manuela. 2016. "Economics Imperialism in Social Epistemology: A Critical Assessment." *Philosophy of the Social Sciences* 46 (5): 443–72.

Gieryn, Thomas F. 1983. "Boundary-Work and the Demarcation of Science from Non-Science: Strains and Interests in Professional Ideologies of Scientists." *American Sociological Review* 48, no. 6: 781–95.

Haber, Ludwig Fritz. 1958. *The Chemical Industry During the Nineteenth Century: A Study of the Economic Aspect of Applied Chemistry in Europe and North America*. Oxford: Clarendon Press.

Hawk, Mark A., Cassandra L. Gorsuch, Patrick Fagan, Chan Lee, Sung Eun Kim, Jens C. Hamann, Joshua A. Mason, et al. 2018. "RIPK1-Mediated Induction of Mitophagy Compromises the Viability of Extracellular-Matrix-Detached Cells." *Nature Cell Biology* 20, no. 3: 272.

Hicks, Daniel J. 2014. "A New Direction for Science and Values." *Synthese* 191, no. 14: 3271–295.

Hicks, Daniel J., Carl Stahmer, and MacKenzie Smith. 2018. "Impacting Capabilities: A Conceptual Framework for the Social Value of Research." *Frontiers in Research Metrics and Analytics* 3.

Hicks, Daniel J., and Thomas A. Stapleford. 2016. "The Virtues of Scientific Practice: MacIntyre, Virtue Ethics, and the Historiography of Science." *Isis: Journal of the History of Science Society* 107, no. 3: 449–72.

Holman, Bennett, and Kevin C. Elliott. 2018. "The Promise and Perils of Industry-Funded Science." *Philosophy Compass* 13, no. 11: e12544.

Hursthouse, Rosalind. 1999. *On Virtue Ethics*. Oxford: Oxford University Press.

Jindal-Snape, Divya, and Jonathan B. Snape. 2006. "Motivation of Scientists in a Government Research Institute." *Management Decision* 44, no. 10: 1325–1343.

Jordan, Catherine, Susan Gust, and Naomi Scheman. 2011. "The Trustworthiness of Research: The Paradigm of Community-Based Research." In *Shifting Ground: Knowledge and Reality, Transgression and Trustworthiness*, edited by Naomi Scheman, 170–90. New York: Oxford University Press.

Kaiser, David. 2004. "The Postwar Suburbanization of American Physics." *American Quarterly* 56, no. 4: 851–88.

Kaiser, David. 2005. *Pedagogy and the Practice of Science: Historical and Contemporary Perspectives*. Cambridge, MA: MIT Press.

Keller, Evelyn Fox. 1983. *A Feeling for the Organism: The Life and Work of Barbara McClintock*. San Francisco: W. H. Freeman.

Knight, Kelvin. 2008. "Practices: The Aristotelian Concept." *Analyse & Kritik: Zeitschrift Für Sozialtheorie* 30, no. 2: 317–29.

Kristjánsson, Kristján. 2018. *Virtuous Emotions*. Oxford: Oxford University Press.

Kuhn, Thomas. 1996. *The Structure of Scientific Revolutions*. Chicago: University of Chicago Press.

Ladner, Joyce. 2002. "Introduction to Tomorrow's Tomorrow: The Black Woman." In *The Gender of Science*, edited by Janet A. Kourany, 353–60. Upper Saddle River, NJ: Prentice Hall.

Lam, Alice. 2011. "What Motivates Academic Scientists to Engage in Research Commercialization: 'Gold,' 'Ribbon' or 'Puzzle'?" *Research Policy* 40, no. 10: 1354–68.

Latour, Bruno, and Steve Woolgar. 1979. *Laboratory Life: The Social Construction of Scientific Facts*. London: Sage Publications.

Liska, Adam. 2005. "The Myth and the Meaning of Science as a Vocation." *Ultimate Reality and Meaning* 28, no. 2: 149–64.

Lucier, Paul. 2010. *Scientists and Swindlers: Consulting on Coal and Oil in America, 1820–1890*. Baltimore: Johns Hopkins University Press.

MacIntyre, Alasdair C. 1984. *After Virtue: A Study in Moral Theory*. Notre Dame, IN: University of Notre Dame Press.

MacIntyre, Alasdair C. 1988. *Whose Justice? Which Rationality?* Notre Dame, IN: University of Notre Dame Press.

Masala, Alberto, and Jonathan Webber, eds. 2016. *From Personality to Virtue: Essays on the Philosophy of Character*. Oxford: Oxford University Press.

Mirowski, Philip, and Esther-Mirjam Sent, eds. 2002. *Science Bought and Sold: Essays in the Economics of Science*. Chicago: University of Chicago Press.

Narvaez, Darcia, and Daniel K. Lapsley. 2005. "The Psychological Foundations of Everyday Morality and Moral Expertise." In *Character Psychology and Character Education*, edited by Daniel K. Lapsley and F. Clark Power, 140–65. Notre Dame, IN: University of Notre Dame Press.

Nicolini, Davide. 2013. *Practice Theory, Work, and Organization: An Introduction*. Oxford: Oxford University Press.

Pew Research Center. 2015. "Public and Scientists Views on Science and Society." https://www.pewinternet.org/wp-content/uploads/sites/9/2015/01/PI_ScienceandSociety_Report_012915.pdf.

Rouse, Joseph. 2007. "Social Practices and Normativity." *Philosophy of the Social Sciences* 37, no. 1: 46–56.

Russell, Daniel C. 2009. *Practical Intelligence and the Virtues*. Oxford: Oxford University Press.

Russell, Daniel C. 2014. "Aristotle on Cultivating Virtue." In *Cultivating Virtue: Perspectives from Philosophy, Theology, and Psychology*, edited by Nancy E. Snow, 17–48. Oxford: Oxford University Press.

Shapin, Steven. 2009. *The Scientific Life: A Moral History of a Late Modern Vocation*. Chicago: University of Chicago Press.

Shapin, Steven. 2010. *Never Pure: Historical Studies of Science as if It Was Produced by People with Bodies, Situated in Time, Space, Culture, and Society, and Struggling for Credibility and Authority*. Baltimore: Johns Hopkins University Press.

Smith, Crosbie. 1998. *The Science of Energy: A Cultural History of Energy Physics in Victorian Britain*. Chicago: University of Chicago Press.

Snow, Nancy E. 2010. *Virtue as Social Intelligence: An Empirically Grounded Theory*. New York: Routledge.

Snow, Nancy E. 2014. *Cultivating Virtue: Perspectives from Philosophy, Theology, and Psychology*. New York: Oxford University Press.

Snow, Nancy E. 2016. "How Habits Make Us Virtuous." In *Developing the Virtues: Integrating Perspectives*, edited by Julia Annas, Darcia Narvaez, and Nancy Snow, 135–56. Oxford: Oxford University Press.

Steel, Daniel. 2017. "Qualified Epistemic Priority: Comparing Two Approaches to Values in Science." In *Current Controversies in Values and Science*, edited by Kevin C. Elliott and Daniel Steel, 49–63. New York: Routledge.

Traweek, Sharon. 1992. *Beamtimes and Lifetimes: The World of High-Energy Physics*. Cambridge, MA: Harvard University Press.

Warren, Mark E. 1999. "What Is Political?" *Journal of Theoretical Politics* 11, no. 2: 207–31.

Warwick, Andrew. 2003. *Masters of Theory: Cambridge and the Rise of Mathematical Physics*. Chicago: University of Chicago Press.

Weber, Max. 1946. "Science as a Vocation." In *From Max Weber: Essays in Sociology*, edited by H. H. Gerth and C. Wright Mills, 129–56. New York: Oxford University Press.

Wittgenstein, Ludwig. 1963. *Philosophical Investigations*. Translated by G. E. M. Anscombe. Oxford: Blackwell.

Zimmerman, Michael J., and Ben Bradley. 2019. "Intrinsic vs. Extrinsic Value." In *The Stanford Encyclopedia of Philosophy*, edited by Edward N. Zalta, Spring 2019. https://plato.stanford.edu/archives/spr2019/entries/value-intrinsic-extrinsic/.

3

Studying Scientific Virtues

Bridging Philosophy and Social Science*

Robert T. Pennock and Jon D. Miller

Studying the Scientific Character Virtues

Philosophers of science have tended to focus on philosophical issues involving scientific concepts in particular fields, such as physics or biology, and scientific reasoning in general, such as the nature of scientific explanation and the justification of inductive inference. Comparatively little attention has been paid to the character of the scientist and the identification and investigation of traits that make for excellence in scientific practice, especially when these are conceptualized as having a moral structure. One strand of Pennock's philosophy of science research has focused on scientific methodology, including confirmation, methodological naturalism, and model-based reasoning; a second strand of empirical research considers the evolution of intelligent behavior. These two strands have been linked by an interest in the relationship of epistemic and ethical values in science, which led him to turn to questions about character and the scientific mindset and to develop an account of scientific excellence and its development that combines Aristotelian virtue ethics and Darwinian evolution (e.g., Pennock 2002, 2015, 2019).

When virtue ethics has been applied to science previously, it has mostly been in terms of how general human virtues may be relevant in science. Pennock's alternative approach, which starts with a narrower notion of *vocational virtues*, however, focuses on character traits as they relate directly to the central, guiding purpose of different vocations. For the practice of science, this approach suggests that there are particular virtues, from veracity and curiosity to objectivity and skepticism, that should be given special weight for fostering excellence in scientific practice and scientific flourishing generally. Over the last couple of decades, Pennock has developed this view philosophically and deployed it for various applications (e.g., Pennock 1996, 2001, 2018; Pennock and O'Rourke 2017). For the most part, it would have been possible to pursue this line of research entirely with a philosophical methodology, but adding a social science component was always an enticing possibility.

Robert T. Pennock and Jon D. Miller, *Studying Scientific Virtues* In: *Science, Technology, and Virtues*. Edited by: Emanuele Ratti and Thomas A. Stapleford, Oxford University Press. © Oxford University Press 2021. DOI: 10.1093/oso/9780190081713.003.0004

From the point of view of sociology of science, an empirical study would provide a look into the culture of science that complemented and extended Robert Merton's work, which had focused on institutional values, setting aside more specific questions about ethical perception and motivation of scientists themselves. And from the point of view of philosophy of science and ethics, such an interdisciplinary study of the ethical perceptions of exemplary scientists held the promise of providing a kind of test of some aspects of this virtue philosophy of science. This kind of investigation required collaboration with a social scientist, which led to the research partnership with Miller and a study of scientists' views about the virtues of their vocation that is now in its sixth year. In this chapter, we will reflect upon some of what we have learned in attempting to bridge these two disciplines and the unique value that doing so has for understanding a cross-cutting topic such as scientific virtue. Before describing our own work, however, we present a couple of cases to show why collaboration between philosophy and social science is infrequent and what might account for the general separation between these disciplines.

When Philosophers and Social Scientists Collaborate

Psychology and the other social sciences began as branches of philosophy, but since the mid-nineteenth century they have gone their own ways, and practitioners in these fields for the most part now operate independently of one another. When philosophers and social scientists do meet, as occurs periodically for scholars in neighboring areas of research, relations are cordial, though their differences can sometimes make it difficult for them to work together on interdisciplinary projects in substantive ways. It is common to ascribe such difficulties to differences in their language. Indeed, in their conversations at interdisciplinary research conferences that they might both attend it does seem at times that scholars are talking past each other. Sometimes they use the same terms but in different ways, which can make a conversation especially confusing. However, their differences are not solely a matter of language; their methods and practices are mostly distinct as well.

Consider the role of data in their respective research about ethical norms. It is not just that ethicists rarely speak of data; they often appeal to ethical intuitions in their arguments, but they do not usually think of these as data and only occasionally use data in a way that social scientists would recognize. If data do appear as part of an ethical argument, they are mostly just part of some background secondary premise and so are taken for granted, as in a utilitarian argument. Why? Because philosophers are interested in figuring out what is proper or appropriate, and this kind of normative question is investigated more commonly by means of analysis and conceptual argument than by empirical observations.

Similarly, it is not just that social scientists rarely talk about values. They may investigate social norms, but not in the evaluative way that philosophers deal with what they call norms. Why? Because social scientists are more interested in describing values than evaluating them. Most social scientists insist upon being scrupulously neutral with regard to different values they may be describing—for example, values from different cultures. It is not their place or interest to make judgments about the values of the people they study.

That ethicists and social scientists' interests are mostly different means that they have little reason to pay attention to what is happening in the other's domain. Their professional literature rarely has the need to cite work from the other, so philosophers are mostly unfamiliar with the social science journals and social scientists rarely dip into the philosophical literature. With their eyes on their own acreage, workers on both sides of the disciplinary wall are thus often unaware of internal conflicts going on in the neighboring field. Ethicists usually have little reason to concern themselves, for instance, with differences between cognitive schema theorists and trait theorists or other developmental camps in psychology, if they notice them at all. Psychologists, similarly, for the most part have little knowledge of, or interest in, debates among ethical theorists about, say, the pros and cons of virtue ethics vis-à-vis deontology or consequentialism.

One finds a similar pattern as one moves from psychology to other social sciences. This is not to say that social scientists have been entirely uninterested in ethics, including the ethics of scientific research itself. However, in line with the differences already noted, the most common sort of studies conducted have been ones that investigate scientific misconduct and attempt to ascertain how common various kinds of questionable research conduct are (e.g., John, Loewenstein, and Prelec 2012). This is still a descriptive question, though in this case a prior specification of what kind of behavior is unethical in science is required. For such a study to get off the ground, social scientists must work with ethicists, at least up front. This is perhaps the simplest form of collaboration—research ethicists say what behaviors are unethical and social scientists then can get on with their business of measuring the incidence of that behavior. For many purposes, this basic form of multidisciplinary collaboration is more than sufficient, as it can be clearly beneficial to get real data about the prevalence of various forms of misconduct. For this kind of research, social scientists need not concern themselves about the ethical theory or reasoning behind the identification so long as they have some well-defined behavior to measure.

As in this example, relatively few social scientists need to or care to delve further into ethical thought itself. Robert Merton's pioneering work on the ethos of science is a notable exception, but even Merton paid little attention to

philosophical issues and mostly discussed the scientific ethos by simply identifying institutional norms and then analyzing their utility for maintaining social
cohesion and behavioral compliance (Merton 1938, 1973). On occasion, however, there have been scholars who have embarked on an investigation on the
border that took them beyond the current division and back to a place where the
disciplines shared some common interests.

Lawrence Kohlberg, who brought investigation of morals back into psychology, is one notable example (Kohlberg 1969). Using a Piagetian model of
developmental stages, Kohlberg studied how people reasoned about moral issues, and in his data he identified six stages (grouped into three levels) that individuals pass through in the development of their moral reasoning. In the earliest,
"pre-conventional" level, children first reason in terms of obedience and punishment (thinking of how to avoid punishment), followed by a self-interest orientation (thinking in terms of "what is in it for me" and how to receive rewards).
Beginning in adolescence, a second level of "conventional" moral reasoning
develops, first with an emphasis on trying to conform to social expectations, and
later with a "law and order" mentality in terms of societal authority and social
conventions. In the "post-conventional" level, more abstract reasoning develops,
first in terms of a notion of social contract, whereby laws for moral behavior are
determined by some joint process for decision for the common good, and then,
in the sixth stage, in terms of universal ethical principles, especially of rights and
justice.

Kohlberg's moral stage theory in the psychology of moral development has
been influential, but its presentation emphasized certain ethical views over
others in ways that some saw as problematic. The highest, sixth stage seemed
to place deontological ethics (exemplified by Kant's view of morals as universal
laws) over utilitarian ethics, which were placed earlier, in the fifth stage. Carol
Gilligan's *In a Different Voice* (1982) suggested that it overlooked a distinctively
female ethics of care. (We will look at that criticism in a bit more detail shortly.)
There was certainly a blind spot with respect to virtue ethics, as pointed out by
David Carr, who thought that was all the worse for the whole Kohlbergian approach, especially with regard to implications for moral education (Carr 1996).
Kohlberg later adjusted and expanded his model in response to some of these
points. When one is researching a topic such as moral psychology, it becomes
necessary to talk with one's neighbors.

In a paper about their own work on "psychologized morality," partially inspired by Kohlberg's research, the psychologists Daniel Lapsley and Darcia
Narvaez reflect on some of the issues that arise for workers at the border between
psychology and ethics (Lapsley and Narvaez 2008). We would like to now add
to their useful reflections for a slightly broader notion of social science, and we
will adopt their use of the metaphor of collaboratively repairing the wall between

these fields from Robert Frost's poem "Mending Wall" (Frost 1914) as a way to frame the issue of the nature of interdisciplinary collaborations of this sort. There is much that could be said about the relationship between these two neighbors, but we want to make just a few points about the relationship between data and values in such collaborations.

Theory: At the Descriptive/Prescriptive Divide

A standard theoretical reason for the separation between ethics and science involves the problematic conceptual gap that philosopher David Hume identified between facts and values: that something is the case does not tell us that it ought to be (Hume 1739). The philosopher G. E. Moore called one form of this leap from the former to the latter (in particular the reduction of a moral term such as "the good" to some natural property such as the pleasurable) the "naturalistic fallacy" (Moore 1903). It is easy to see how this problem applies to the sort of interdisciplinary research we are considering. Social science investigates factual questions and provides descriptive information. Ethical philosophy investigates value questions and provides prescriptive evaluations. Both sorts of inquiries are sometimes referred to as "normative," which adds confusion to what is already a notoriously complex issue, but one basic way to put the point is as follows: No matter what a descriptive study finds about community C's or subject S's ethical beliefs, ethics may still properly ask whether C or S should or should not hold those beliefs. Thinking that mere facts about a culture's or subject's views are sufficient to determine ethics leads to the conceptual incoherence of cultural relativism or subjectivism. This is Philosophy 101. Kohlberg took this issue seriously and wrote a major article about how his empirical work could avoid the pitfalls of the naturalistic fallacy (Kohlberg 1971).

This distinction is the most plausible reason why philosophers and social scientists are happy to stay in their respective fields with a wall that marks a dividing line between their areas of interest. But what is the nature of this wall? If it is taken to be an impermeable barrier, then one might wonder what use information gathered from science generally and social science in particular could ever have for ethics. On the other hand, if it is taken to mark a strict separation, then it might serve as a regulative principle for both philosophers and social scientists, making it clear when one has improperly trespassed on the other's property. We may gain some insight into the nature of the wall by looking at a couple of cases of border crossings that were occasion for both friendly and more strained interactions between the neighbors.

An especially interesting border case involved issues raised by Kohlberg's research. His own work on stages of moral development became a topic of discussion in philosophy courses, but it did not have nearly the impact as the work of Carol Gilligan, who began as his research assistant but whose own research led her to identify what she purported was a distinctively different, "female" moral voice that reasoned in terms of care for others and preservation of relationships compared to a "male" focus on rights and fairness. Here is a case where empirical data from social science seemed to directly influence ethicists. Gilligan's work led to suggestions for reappraisal of standards of ethical judgment, first among feminist philosophers and then more broadly. Largely because of the influence of her empirical work, it became and remains common to include discussion of care-based ethics in ethics textbooks.

Not all philosophers embraced Gilligan's work wholeheartedly, however. Ethicist Owen Flanagan warned of philosophical dangers he saw in her work and Kohlberg's as well, especially with regard to claims about particular hierarchies of moral adequacy, which presumed resolution to disputes among different ethical theories that had by no means been accomplished. Kohlberg's account drew heavily from the influential work of ethicist John Rawls, but important as that theory of justice was and remains, it seemed premature to crown its notion of morality as more justified than utilitarian or other ethical accounts. Divisive arguments about whether "male" justice or "female" care is superior bear witness to how psychological claims can negatively affect moral discourse. Flanagan recommended that discussion of possible differences between how women and men view morality ought better be pursued "in a scientific context freed of the ideological weight of moral stage theories with their tenuous suite of claims about universality, irreversibility, and, especially, adequacy" (Flanagan 1982, 512).

Philosopher Debra Nails was harsher, arguing on both philosophical and empirical grounds that Gilligan's work was "fallacious and unreliable" (Nails 1983). She was especially critical of reifying psychological scales of "moral maturity," but she also delved into various methodological problems with the interview process, calling into question Gilligan's conclusions about differences between male and female moral standpoints. Subsequent empirical work bore out this criticism. Other researchers were not able to replicate Gilligan's results, and a meta-analysis of research on gender differences in moral orientation showed at best a weak relationship between care and justice orientations in women and men (Jaffee and Hyde 2000). Nails did point out that Gilligan had noted at the very beginning of *In a Different Voice* that the different voice she described was characterized by theme, not gender, but that important caveat was mostly lost in the book itself and in the subsequent reception of the work. Nails concluded with a warning: "This type of research is social science at sea without anchor, and no one is out of danger" (Nails 1983, 664).

Now we turn to a second example—one where the challenge comes from the other direction. This case involves what is known as the "situationist critique" of virtue ethics. We have already mentioned the debate between virtue ethics and theories of deontology and utilitarianism as an example of an internal topic of interest among ethicists. Virtue ethics, which derives from Aristotle, especially his *Nicomachean Ethics*, and has seen a resurgence of interest in recent decades, focuses on the notion of moral character. Rather than dealing primarily with the morality of actions and events as deontological and utilitarian ethics do, virtue ethics deals first with agents and states of agents. In an Aristotelian framework, virtue ethicists analyze how the development of moral character leads to human flourishing. If a person is courageous, for instance, that trait may be expected to motivate appropriately courageous behavior when circumstances warrant it. Similarly, having the virtue of generosity ought to lead one to act generously in appropriate situations. Aristotle described various ways that virtues can be developed, such as by learning from other individuals who have acquired practical wisdom.

Deliberation about the justification of ethical theory mostly involves conceptual arguments, but in an unusual case this standard account of virtue ethics was criticized on empirical grounds using data from psychology. The philosopher Gil Harman, pointing to empirical studies that highlighted the problem of explaining behavior in terms of distinctive characteristics of an agent over relevant details of the agent's perceived situation, called the virtue ethicists' appeal to moral character traits an example of a "fundamental attribution error" (Harman 1998–99, 329–30). Ethicist John Doris extended this critique, highlighting an experiment from social psychologists Alice Isen and Paula Levin (1972) that showed how helping behavior varied as a function of mood state—that is, of "feeling good" in a situation. In their experiments, something as simple as finding a dime in the coin return of a public phone led to a statistically significant increase in helping behavior. Doris cited this experiment as representative of the "situationist" tradition in social and personality psychology, which rejects explanations of behavior in terms of standard theoretical constructions of personality, and he concluded, like Harman, that it calls into question neo-Aristotelian virtue ethics based on the idea of steady character traits (Doris 1998, 504). Such experiments suggest, he argued, that one cannot reliably predict helping behavior based on traits such as selfishness or compassion (as in his hypothetical example of subjects in the dime study) and that an alternative conception of moral personality and of ethics is required that does not posit reliable moral dispositions.

The descriptive/normative divide notwithstanding, this situationist challenge was recognized by ethicists as a real issue for neo-Aristotelian virtue ethics. One may of course say, as Doris himself acknowledged, that showing the descriptive inadequacy of an ethical theory is not sufficient for showing that it is normatively inadequate, but the descriptive data do bear upon its plausibility, which led to a lively debate in the literature. We mention just a couple of examples.

Diana Fleming confronted the situationist challenge from both sides, questioning the empirical evidence and its relevance. Experiments such as Isen and Levin's, she pointed out, are "poorly suited to detecting the existence or influence of traits" (Fleming 2006, 38), as they focus on particular situations rather than behavioral consistency over long periods of time and thus "bias the evidence in favor of situational factors" (38–39). She also rightly pointed out that it was a fundamental mistake to think that character traits should operate "without regard to environmental factors," and she argued that once this is acknowledged, it is easy to see that virtue ethics has the conceptual resources to build in "situation sensitive" traits (41).

Nancy Snow responded to the situationist challenge in even greater detail in a book-length account that defended an account of virtue as "social intelligence" (Snow 2010), with extensive discussion of social science research to ground her theory. Contra arguments that situationist experiments undermine the notion of stable character traits, she pointed to other social science research showing how agents' responses to situations also depend upon how they construe the situations they find themselves in. This adds another dimension to moral behavior; it depends not just on dispositions and environments but also on construal. In addition to mood-effect studies such as those of Isen and Levin, Snow addressed a range of others including the infamous Stanford Prison Experiment and the Milgram obedience-to-authority experiments, concluding that situations have meaning for people and that subjects' moral behaviors must be judged with that factor in mind (Snow 2010, ch. 5).

It is not our intention to get into the details or take a stand on the specifics of these debates. We outline them here for the way they reveal the various sorts of conversations that may arise at the border between philosophy and social science. They also illustrate occasional conflict between these parties, though that need not be inevitable; confusions and potential conflicts may be sorted out if each side brings its own expertise to the conversation and does not presume to discount that of the other.

Kohlberg may have neglected virtue ethics, but we are not as pessimistic as Carr in thinking that any such view is incompatible with virtue ethics. The idea of a universal morality surely does not preclude the place of moral judgment or the goal of learning from the practically wise how to balance complex values when considering what is ethical in particular cases. Moreover, even though the empirical basis of Gilligan's claim of a distinctively female ethical voice seems to have been mostly undermined, her idea of an ethics of care remains of philosophical interest. Future research may shed light on whether there is any psychological import to the small empirical differences between men and women with regard to care-based versus justice-based moral reasoning, but no matter how this turns out, ethicists could still argue about how to balance justice and care in moral reasoning and decision-making.

Situationists may be wrong in their specific challenge, or at least premature in their conclusions, but they could have been right, and there are other empirical data that may be relevant. For instance, if moral psychology were to show that human beings are psychologically incapable of being virtuous or behaving virtuously, then that would seriously reduce the viability of applying virtue ethics to human morality. Metaethics correctly notes that ought implies can, which means that we at least need to be capable of virtue if it is to be recommended as the correct moral account for human beings. It is extremely unlikely that we are incapable of virtue, but the same point holds for weaker possibilities. The general point for us here is that the entire debate occurs because everyone agreed that for some important aspects of virtue ethics, the empirical data from social science is relevant. As Snow rightly noted, "Philosophers have much to learn from studying the work of psychologists" (Snow 2010, 117).

What these cases show is that there are interesting and substantive points of contact between descriptive psychology and prescriptive virtue ethics where each may reasonably influence the other. These and similar debates show that social science and ethics are not entirely independent and that interesting things may happen where the fields meet. There is much to talk about over the wall, including questions about the wall's proper placement.

When conflict arises between philosophical and social science perspectives, it could easily turn into a war or at least a standoff if one thought that the issue involves which side is more fundamental and gets to decide the answer. But this is usually the wrong way to think about such cases, as the sides are mostly interested in different questions, as we saw, and conflicts are mostly the result of miscommunication about their aims. Clarifying not just terms but methodological assumptions can go a long way toward maintaining peaceful relationships at the border. As we have also seen, however, in other cases there may be real points of contact where some questions do touch substantively on both sides. In such instances, neighborly relations require a more cooperative approach.

While it may offer data about moral development, social science is worth taking seriously only if it has constructed its instruments and protocols so that we can trust that what they are measuring is ethically significant, rather than some related but non-moral trait. Are moral psychologists (and other social scientists) really telling us something about virtue? Any study needs to begin with the appropriate theory (here an ethical theory) before it can state meaningful hypotheses to test. This requires the expertise of philosophers. For instance, an ethicist would need to first explain what being courageous means in virtue-ethical terms, say, as opposed to what it means colloquially, before psychological studies can investigate how to cultivate it as a moral trait. A study that failed to appreciate the difference between the philosophical notion of a character trait and the psychological notion of a personality trait is likely to confuse rather than enlighten.

In the other direction, when studies are appropriately relevant, ethicists must be constrained by what social science finds out about limits of human ethical behavior, or other facts that bear upon one or another aspect of ethical thought and behavior. In such cases, data may make a real difference and could even play a role in assessing different ethical theories. For example, virtue ethics claims an advantage over other ethical theories in that moral motivation is supposedly built into its view. It could turn out, however, that social science finds that ethical motivation works differently than theory predicts—perhaps, for instance, training in a rights-based theory actually is better than training in a virtue-based approach at leading to moral action. Again, this is a hypothetical scenario, but if empirical data substantiated that possibility, then it would be incumbent upon ethicists to consider its implications and rethink their views.

These are just a few examples where collaboration between ethicists and social scientists could prove to be fruitful for both. We now return to our own collaboration, which we believe is another case where an interdisciplinary approach opens up new ways of thinking about some old questions.

Our Scientific Virtues Survey Collaboration

Pennock's philosophical theory of vocational virtue holds that practices and methods that characterize a discipline are justified in relation to its central, guiding purpose and general human values, and that the character virtues involved in the excellence and integrity of that vocation are those traits that one ought to cultivate for it to flourish. With regard to science, which aims to discover truths about the natural world, the theory thus holds that scientists ought to value traits such as veracity, curiosity, objectivity, and related virtues; these would be expected to constitute the scientific mindset, helping science flourish. To see what ideals scientists themselves actually held, in the late 1990s Pennock created a preliminary questionnaire and began a series of informal interviews with scientists. This provided anecdotal support for the account, but it wasn't until the early 2010s that the opportunity arose to develop and mount a systematic, formal, national study in collaboration with Miller.

Miller's work has focused on empirical descriptions of adolescent and adult development of life plans and goals, including occupational objectives and the acquisition of skills and information needed to achieve various life goals (e.g., Miller 1983, 2000, 2010). This life course approach is reflected in two major strands of work.

In 1987, Miller launched the Longitudinal Study of American Youth (LSAY) to study the development of student engagement with science and mathematics and the impact of these early learning experiences on the selection of career paths and life goals.[1] The study adopted Berger and Luckman's developmental perspective (Berger and Luckman 1966), which minimizes genetic explanations and argues that newborn children are blank slates that are shaped by their family and life course experiences. Language plays a central role in the definition of reality and the transmission of values. In this model, a child born in Ann Arbor, Michigan, and transported to Beijing within a few days of birth and raised by a Chinese family would be expected to grow up speaking Mandarin and to adopt the values and norms of his or her significant others (the adults who nurture and raise the child). Conversely, a child born in Beijing and transported to Ann Arbor within the first few days of life and raised by a midwestern American family would grow up speaking American English and would adopt the values and norms of his or her American significant others.

Applying this model, the LSAY collected questionnaires and achievement tests for its 5,900 participating students several times each year and extensive questionnaire information from each student's science and mathematics teachers. One parent of each child was interviewed by telephone each year during secondary school. Extensive school and community context information was added to each student's record. After secondary school, the LSAY followed each student to work or postsecondary study, using a combination of mailed questionnaires and telephone interviews. In 2007, the LSAY changed to a system of online data collection for all respondents who were willing to respond online and to mailed print questionnaires for other respondents. The LSAY continues to follow 5,100 of the original 5,900 respondents (the eligible sample who are still alive, living in the United States, and mentally and physically able to complete a questionnaire).

The second strand of Miller's work is a forty-year time series of adult studies that measure patterns of adult issue interest, scientific literacy, and democratic engagement in the formulation of science and technology policy in the United States. The insights gained from the developmental analysis in the LSAY allow the construction of questions for cross-sectional respondents that provide useful development indicators (obtained by recall). But the selection of a new probability sample of American adults each year allows the construction of time trends over decades.

At one level, all of this work is empirical and descriptive, but it also reflects a commitment to democracy and democratic values. A good deal of Miller's work has been based on his measurement of civic scientific literacy—defined as the level of scientific construct understanding necessary for a citizen to engage in science and technology policy discussions. Democracy depends on meaningful citizen participation, which implies a level of issue understanding sufficient to make policy-relevant judgments in voting and in contracting or pressuring electing representatives. A second set of value issues pertain to educational and social disparities in society. One of the results of consistent measurement over

STUDYING SCIENTIFIC VIRTUES 69

a period of decades is that it is possible to see patterns of differentiation among individuals and groups. Miller's work has focused attention on the growing disparity in educational opportunity and attainment in the United States and its consequences for employment, health, and quality of life. Here the distinction between description and prescription does apply, as the data do not by themselves prescribe a specific solution or even a precise definition of a problem. However, they do provide a common ground for discussion.

Both strands of this social science research were relevant to collaborating on an interdisciplinary investigation of scientific character virtues and values. We began our study in 2012 with pilot interviews to develop our survey instrument and kicked off the full national study in 2014. Drawing a random sample from the population of peer-identified exemplary scientists (e.g., elected fellows of major scientific societies) and a population of early-career scientists, we then solicited and conducted phone interviews, which averaged forty-five minutes. Subjects who did not want to do an interview could provide written answers. By 2018 we had data from more than 1,100 scientists.

There are many things that one could say about the nature of this collaboration, but this is not the place to discuss the details of the protocol or the results. Here we just want to highlight three points about our study that are illustrative of some of the methodological and other issues that we think are of interest for philosophers and social scientists who walk the wall at the border between their fields.

First, the core of our survey is not about what virtues scientists have but what ones they most value. This helped us get at the questions that are of philosophical interest. We did not ask about what character virtues/traits scientists do or do not exhibit; rather, we asked what traits were important (i.e., valuable) for science. This is asking for their normative assessment (i.e., their evaluative, ethical judgment) of possible traits.

In this way, our survey has a prescriptive focus on what scientists think they should aspire to, rather than a descriptive focus on what traits they may or may not display. It would be a very different kind of study to investigate how scientists are actually behaving. That kind of descriptive information is useful for other purposes, of course. Other studies have tried to investigate how widespread scientific misconduct (for instance, fabricating data) is, but for us the question of how many scientists depart from the virtue of honesty is secondary to the question of how important they think that value is to science in the first place. For many social scientists, this may not be a distinction that makes much difference for their purposes—it is just one measure of attitude among any number of others. But for philosophers, ideals and aspirations are normative in more than a descriptive sense, as they take us into the realm of the asking and giving of reasons. This is grist for the philosophical mill.

There are other aspects of our study, however, that are familiar to social scientists but will seem foreign to most ethicists. We turn now to two related examples.

For one set of questions in our instrument we asked subjects to give a quantitative rating of a list of virtues. For each virtue, we asked them to assess its importance in science by rating how essential the trait was to being an exemplary scientist on a scale of 0 to 10, with 0 being not at all essential and 10 meaning that it is very essential. This is not the sort of thing that ethicists would typically care or think to ask. Quantitative assessments do come into play in utilitarian arguments, where doing the right thing (i.e., that which produces the greatest good for the greatest number of people) requires that we assess the effects on happiness (or pleasure) that different choices would have. In reality, however, it is more likely to be economists who actually collect and analyze such data; utilitarian ethicists work out the theory but do not usually themselves apply it in this way. So, besides that, what possible philosophical value is there in asking for a quantitative assessment of the importance of some value/virtue?

One way to begin to answer this is to consider the alternative. For instance, take any random virtue and one could likely find a way that it could be useful in some particular scientific investigation. Take generosity, for example. An advocate for this virtue could easily find or concoct a case where it would behoove scientists to be generous—in the sharing of samples, for instance. If the only question were whether a virtue could be relevant for a scientist, one could come up with possible scenarios without troubling to run a survey. But this is where data can be revealing: it would be interesting to find that generosity (or some other common virtue) does not register as very important for scientists (or that some uncommon virtue does). We may certainly admit that generosity is a general virtue for human society without thinking that it should be applied with equal force in all human vocations compared to other virtues. This is why moral judgment is important after all.

Quantitative information from exemplary scientists (or some other vocational practitioners) provides a broad measure of the judgments of practiced researchers—those who presumably have acquired some practical wisdom from their experience. This kind of data is surely relevant to the assessment of the claims of virtue ethics, and it is better than a simple appeal to the intuitions of one or a few ethicists.

Moreover, such quantitative data allows one to ascertain whether there are patterns of emphasis. It could turn out, for example, that exemplary practitioners of other vocations all agree that everything on our list, or some other, were indeed virtues (i.e., that it is generally a good thing to be honest, curious, objective, and so on, or generous, caring, and chaste), which is by no means certain, but that would not be the end of the matter. If one asked only whether something was a virtue or not, one might have to leave it at that, but quantitative ratings allow one to see whether there are patterns of weights that practitioners of different vocations give to different virtues. Pennock's vocational virtue account predicts that there ought to be such differences, depending in large part on what is the central, guiding purpose of the practice. If exemplary practitioners of different vocations all gave the same weights to the same set of virtues, then that would be a good indication that the account is wrong.

Of course, we did not stop with numerical ratings; we also collected a wide range of qualitative information. We asked scientists to explain the reasoning for their highest scores. We asked them to describe examples of how these appeared in practice. We asked them how they learned about these values and how they tried to pass them on to their own students. We asked them about scientific vices. We asked for anecdotes and stories that illustrated their views. Such qualitative data will be interesting and useful for understanding the nuances of these views and the reasoning behind them. Philosophers rarely collect such information in a systematic fashion; their method relies more on conceptual analysis, where a single case—sometimes even a hypothetical case—can provide fodder for extended reflection, discussion, and debate. Imagined examples can be as powerful as real ones when one is trying to work out how to think about what we ought or ought not do in difficult ethical dilemmas, so it may not matter whether the example was drawn from life or not. Because philosophical thought experiments are the bread and butter of much philosophical deliberation, many philosophers will have little use for even this sort of qualitative empirical data. However, it is the quantitative data and how to incorporate it into philosophical analysis that will require more significant rethinking of their usual modes of reasoning. This brings us to a third point of methodological interest that involves a specific, technical method of quantitative reasoning that we used.

Confirmatory factor analysis (CFA) is common in the social sciences, but it is probably fair to say that our study is the first time that it has been part of a philosophical investigation. The collection of the relative importance of each of several virtues provides a useful empirical basis for analyzing whether they cluster or are likely to be associated. A CFA procedure uses the variance or covariance among a set of variables (our measures of the importance of each virtue) and asks whether all of the measured virtues are of equal importance or whether some are distinctively more important and more likely to be selected together by exemplary scientists. In drawing descriptive conclusions, social scientists pay attention to critical methodological principles from the philosophy of science about the importance of subjecting hypotheses to rigorous tests, and CFA is one way to do this for certain kinds of models.

We see this examination of whether items (virtues) cluster or covary as a test of one aspect of the hypothesized view of scientists' values. If we start with the premise that all of the virtues included in our final questionnaire (which were selected through pilot testing of a much larger number of prospective virtues) are equally likely to occur in a final cluster of virtues, then we would expect a uniformly strong (or high) set of factor loadings. But we found that only eight of the ten virtues we asked about in the final questionnaire formed a cohesive cluster and that two did not. We would characterize this finding as disconfirming of the central role of the two virtues that did not load on the common factor.

This finding does not mean that the two virtues that did not load on the common factor for our sample of exemplary scientists would not load on a similar factor for another sample of exemplary scientists or for similar samples of engineers, bankers, or other professionals. Neither does it mean that these virtues may not be important in some other way or for a subset of research. For instance, we hypothesize that the virtue of being collaborative did not load together with the others because subjects thought of significant cases of exemplary science that were the work of a scientist mostly working independently.

If hypotheses or expectations are disconfirmed in a number of examinations of this kind, scientists and social scientists would conclude that the virtues in question do not belong as a part of a core set of virtues. Ethicists should take heed as well, for this may be revealing of some unrecognized feature of the moral landscape of science.

Conclusion

In Frost's poem, two neighbors meet in the spring to repair their common wall. One notes that a wall is not really needed, as "he is all pine and I am all apple orchard," but the other says that good fences make good neighbors, so together they close any gaps that have formed over the winter. In their reflection on the neighboring fields of ethics and developmental psychology, Lapsley and Narvaez suggested that a fence is good "not because it keeps neighbors apart and distant but because the occasion of repairing it gives them reason to work together in partnership" (Lapsley and Narvaez 2008, 280). In this chapter we have examined a few elements that are relevant to that common task when ethics is considered in relation to the somewhat broader field of social science, focusing especially on the relevance of empirical data.

As we have seen, philosophers and social scientists do have different primary interests and these differences of interest may sometimes lead to their talking past one another, as is often the case for many interdisciplinary conversations. However, many such problems can be avoided once the two groups recognize and articulate the relevant differences. Difference of interests should be acknowledged (that is what the mending of the wall involves) so that neither side inadvertently insists that only their interests are legitimate (it would be a mistake to root out the apple trees to plant more pines, or vice versa).

We have also identified a few ways in which each neighbor's interest may properly intersect with the other's; where data and values may meet; and where practitioners on both sides may find their perspectives broadened by interaction with the other. In such cases, philosophers and social scientists may collaborate for their mutual benefit. Putting the point poetically, we may conclude that in mending walls between fields of apples and pine, it is perhaps worth considering the value of leaving a gap, perhaps with a gate, that will allow a way for the neighbors to cross over and be wide enough so that "even two can pass abreast."

Notes

* This material is based upon work supported by grants to Pennock and Miller by the John Templeton Foundation under Cooperative Agreement Nos. 36630 and 42023, and to Pennock by the National Science Foundation under Cooperative Agreement No. DBI-0939454. Any opinions, findings, and conclusions or recommendations expressed in this material are those of the authors and do not necessarily reflect the views of the John Templeton Foundation or the National Science Foundation.
1. Originally funded by the NSF for twenty years, the study is now funded by the National Institute on Aging and has been renamed the Longitudinal Study of American Life.

References

Berger, Peter L., and Thomas Luckmann. 1966. *The Social Construction of Reality: A Treatise in the Sociology of Knowledge*. New York: Anchor Books.
Carr, David. 1996. "After Kohlberg: Some Implications of an Ethics of Virtue for the Theory of Moral Education and Development." *Studies in Philosophy and Education* 15: 353.
Doris, John M. 1998. "Persons, Situations, and Virtue Ethics." *Noûs* 32, no. 4: 504–30.
Flanagan Jr., Owen J. 1982. "Virtue, Sex, and Gender: Some Philosophical Reflections on the Moral Psychology Debate" *Ethics* 92, no. 3: 499–512.
Fleming, Diana. 2006. "The Character of Virtue: Answering the Situationist Challenge to Virtue Ethics." *Ratio* 19, no. 1: 24–42.
Frost, Robert. 1914. *North of Boston*. New York: Henry Holt and Company.
Gilligan, Carol. 1982. *In a Different Voice: Psychological Theory and Women's Development*. Cambridge, MA: Harvard University Press.
Harman, Gilbert. 1998–99. "Moral Philosophy Meets Social Psychology." *Proceedings of the Aristotelian Society* 99, no. 1: 315–31.
Hume, David. 1739. *A Treatise of Human Nature*. London: John Noon.
Isen, Alice M., and Paula F. Levin. 1972. "Effect of Feeling Good on Helping: Cookies and Kindness." *Journal of Personality and Social Psychology* 21, no. 3: 384–88.
Jaffee, Sara, and Janet Shibley Hyde. "Gender Differences in Moral Orientation: A Meta-Analysis." *Psychological Bulletin* 126, no. 5: 703–72.

John, Leslie K., George Loewenstein, and Drazen Prelec. 2012. "Measuring the Prevalence of Questionable Research Practices with Incentives for Truth Telling." *Psychological Science* 23, no. 5: 524–32.

Kohlberg, Lawrence. 1969. "Stage and Sequence: The Cognitive Development Approach to Socialization." In *Handbook of Socialization Theory*, edited by D. A. Goslin, 347–480. Chicago: Rand McNally.

Kohlberg, Lawrence. 1971. "From Is to Ought: How to Commit the Naturalistic Fallacy and Get Away with It in the Study of Moral Development." In *Cognitive Development and Psychology*, edited by T. Mischel, 151–235. New York: Academic Press.

Lapsley, Daniel K., and Darcia Narvaez. 2008. "'Psychologized Morality' and Ethical Theory, or, Do Good Fences Make Good Neighbors?" In *Getting Involved: Global Citizenship Development and Sources of Moral Values*, edited by Fritz K. Oser and Wiel Veugelers, 279–91. Rotterdam: Sense.

Merton, Robert K. 1938. "Science and the Social Order." *Philosophy of Science* 5: 321–37.

Merton, Robert K. 1973. "The Normative Structure of Science." In *The Sociology of Science: Theoretical and Empirical Investigations*, 267–78. Chicago: University of Chicago Press.

Miller, Jon D. 1983. "Scientific Literacy: A Conceptual and Empirical Review." *Daedalus* 112, no. 2: 29–48.

Miller, Jon D. 2000. "The Development of Civic Scientific Literacy in the United States." In *Science, Technology, and Society: A Sourcebook on Research and Practice*, edited by David D. Kumar and Daryl E. Chubin, 21–47. New York: Plenum.

Miller, Jon D. 2010. "The Conceptualization and Measurement of Civic Scientific Literacy for the 21st Century." In *Science and the Educated American: A Core Component of Liberal Education*, edited by John G. Hildebrand and Jerrold Meinwald, 241–55. Cambridge, MA: American Academy of Arts and Sciences.

Moore, G. E. 1903. *Principia Ethica*. Cambridge: Cambridge University Press.

Nails, Debra. 1983. "Social-Scientific Sexism: Gilligan's Mismeasure of Man." *Social Research* 50, no. 3: 643–64.

Pennock, Robert T. 1996. "Inappropriate Authorship in Collaborative Scientific Research." *Public Affairs Quarterly* 10, no. 4: 379–93.

Pennock, Robert T. 2001. "The Virtuous Scientist Meets the Human Clone." In *New Ethical Challenges in Science and Technology: Sigma Xi Forum 2000 Proceedings*, 117–24. Durham, NC: Sigma Xi.

Pennock, Robert T. 2002. "Research Funding and the Virtue of Scientific Objectivity." *Academic Integrity* 5, no. 2: 3–6.

Pennock, Robert T. 2015. "Fostering a Culture of Scientific Integrity: Legalistic vs. Scientific Virtue-Based Approaches." *Professional Ethics Report* 2: 1–3.

Pennock, Robert T. 2018. "Beyond Research Ethics: How Scientific Virtue Theory Reframes and Extends Responsible Conduct of Research." In *Cultivating Moral Character and Virtue in Professional Practices*, edited by David Carr, 166–77. Abingdon, UK: Routledge.

Pennock, Robert T. 2019. *An Instinct for Truth: Curiosity and the Moral Character of Science*. Cambridge, MA: MIT Press.

Pennock, Robert T., and Michael O'Rourke. 2017. "Developing a Scientific Virtue-Based Approach to Science Ethics Training." *Science and Engineering Ethics* 23, no. 1: 243–62.

Snow, Nancy, E. 2010. *Virtue as Social Intelligence: An Empirically Grounded Theory*. New York: Routledge.

PART II
VIRTUES AND TECHNOLOGY

4

Twenty-First-Century Virtue

Living Well with Emerging Technologies*

Shannon Vallor

Introduction

In May 2014, cosmologist Stephen Hawking, computer scientist Stuart Russell, and physicists Max Tegmark and Frank Wilczek published an open letter in the UK news outlet *The Independent*, sounding the alarm about the grave risks to humanity posed by emerging technologies of artificial intelligence. They invited readers to imagine these technologies "outsmarting financial markets, out-inventing human researchers, out-manipulating human leaders, and developing weapons we cannot even understand" (Hawking et al. 2014). The authors asserted that while the successful creation of artificial intelligence (AI) has the potential to bring "huge benefits" to our world and would undoubtedly be "the biggest event in human history," "it might also be the last." Hawking echoed the warning later that year, telling the BBC that unrestricted AI development "could spell the end of the human race" (Cellan-Jones 2014).

While some high-profile inventors such as Elon Musk, Steve Wozniak, Bill Gates, and others remain concerned about AI's potential to pose an existential threat to meaningful human existence, many leading AI experts dismiss these warnings as baseless fear-mongering. Yet even as overhyped fears about the rise of "robot overlords" and machine forms of "superintelligence" finally begin to cool amid evidence of a far more prosaic path for AI development, the ethical challenge to humanity posed by AI and other emerging twenty-first-century technologies continues to grow. The first wave of commercial AI applications, from facial recognition tools to predictive policing algorithms and new medical diagnostics, has sparked international moral and political debate about how to ensure respect for privacy, social justice, and human rights in societies increasingly shaped by automated algorithmic decision-making. The chorus of voices calling for wiser and more effective human oversight of these new technologies grows louder every day. How worried should we be? More importantly: what should we *do*?

Shannon Vallor, *Twenty-First-Century Virtue* In: *Science, Technology, and Virtues*. Edited by: Emanuele Ratti and Thomas A. Stapleford, Oxford University Press. © Oxford University Press 2021. DOI: 10.1093/oso/9780190081713.003.0005

AI is only one of many emerging developments—from genome editing and climate geoengineering to the globally networked Internet of Things—shaping a future unparalleled in human history in its technological promise *and* its peril. Are we up to the challenge this future presents? If not, how can we get there? How can humans hope to live well in a world made increasingly more complex and unpredictable by emerging technologies? As I have argued in more detail elsewhere (Vallor 2016), in essence my answer is this: we need to cultivate in ourselves, collectively, a special kind of moral character, one that expresses what I will call the *technomoral virtues*.

What do I mean by technomoral virtue? To explain this concept will require introducing some ideas in moral philosophy, the study of ethics. At its most basic, ethics is about what the ancient Greek philosopher Socrates called the "good life": the kind of life that is most worthy of a human being, the kind of life worth choosing from among all the different ways we might live. While there are many kinds of lives worth choosing, most of us would agree that there are also some kinds of lives *not* worth choosing, since we have better alternatives. For example, a life filled mostly with willful ignorance, cruelty, fear, pain, selfishness, and hatred might still have some value, but it would not be a kind of life worth *choosing* for ourselves or our loved ones, since there are far better options available to us—better and more virtuous ways that one can live, for ourselves and everyone around us. But what does ethics or moral philosophy have to do with technology?

Ethics as Technomoral Practice

In reality, human social practices, including our moral practices, have always been intertwined with our technologies.[1] Technological practices—everything from agriculture and masonry to markets and writing—have shaped the social, political, economic, and educational histories of human beings. Today, we depend upon global systems of electronic communication, digital computation, transportation, mass manufacturing, banking, agricultural production, and healthcare so heavily that most of us barely notice the extent to which our daily lives are technologically conditioned. Yet even our earliest ancestors used technology, from handaxes and spears to hammers and needles, and their tools shaped how they dealt with one another—how they divided their labor, shared their resources and living space, and managed their conflicts. Among our primate cousins, female chimpanzees have been observed to stop fights among males through technological disarmament—repeatedly confiscating stones from an aggressor's hand (de Waal 2007, 23).

Ethics and technology are connected because technologies invite or afford specific patterns of thought, behavior, and valuing; they open up new possibilities for human action and foreclose or obscure others. For example, the invention of the bow and arrow afforded us the possibility of killing an animal from a safe distance—or doing the same to an otherwise stronger human rival, something that changed the social and moral landscape. Today's technologies open their own new social and moral possibilities for action. Indeed, human technological activity now reshapes the very planetary conditions that make life possible. In the midst of a sixth mass extinction and accelerating climate instability driven by unchecked human industrial activity, we turn to new technologies—for carbon capture, for renewable energy production, for ocean plastic removal—to preserve a future for ourselves and other species. Thus twenty-first-century decisions about how to live well—that is, about ethics—are not simply moral choices. They are *technomoral* choices, for they depend on the evolving affordances of the technological systems that we rely upon to support and mediate our lives in ways and to degrees never before witnessed.

While ethics has always been embedded in technological contexts, humans have, until very recently, been the primary authors of their moral choices, and the consequences of those choices were usually restricted to impacts on individual or local group welfare. Today, however, our aggregated moral choices in technological contexts routinely impact the well-being of people on the other side of the planet, a staggering number of other species, and whole generations not yet born. Meanwhile, it is increasingly less clear how much of the future moral labor of our species will be performed by human individuals. Driverless cars are already being programmed to make responsible driving decisions on our behalf (even as other cars roll out of the factory programmed to cheat on their innocent owners' emissions tests).[2] High-frequency trading algorithms now direct the global flow of vital goods and wealth at speeds and scales no human analyst can match. Artificially intelligent life coach apps are here to "nudge" us when we need to lower our voices, call our mothers, or write nicer emails to our employees. Advanced algorithms inscrutable to our inspection increasingly do the work of labeling us as combatant or civilian, good loan risk or future deadbeat, likely or unlikely criminal, hireable or unhireable.

For these reasons, a contemporary theory of ethics—that is, a theory of what counts as a good life for human beings—must include an explicit conception of how to live well with technologies, especially those that are still emerging and have yet to become settled, seamlessly embedded features of the human environment. Robotics and artificial intelligence, new social media and communications technologies, digital surveillance, and biomedical enhancement technologies are among the emerging innovations that will radically change the kinds of lives from which humans are able to choose in the twenty-first century and beyond. How can we choose wisely from the apparently endless options that emerging technologies offer? The choices we make will shape the future for our children, our societies, our species, and others who share our planet, in ways never before possible. Are we prepared to choose *well*?

The Problem of Technosocial Opacity

This question involves the future, but what it really asks about is our readiness to make choices in the *present*. The twenty-first century is entering its adolescence, a time of great excitement, confusion, and intense anxiety. As with many adolescents, our era is also deeply self-absorbed. In popular and scholarly media, we find both historical consciousness and the "long view" of humanity giving way to an obsessive quest to define the distinctive identity of the present age, an identity almost always framed in technological terms. Whether we call it the Fourth Industrial Revolution, the Age of Information, the Mobile Era, the New Media Age, or the Robot Age, we seem to think that defining the technological essence of our era will allow us to better fathom the course of its future—*our* future.

Yet in one of those cruel paradoxes of adolescence, all our ruminations and fevered speculations about the mature shape of life in this century seem only to make the picture *more* opaque and unsettled, like a stream bottom kicked up by shuffling feet. Among all the contingencies pondered by philosophers, scientists, novelists, and armchair futurists, the possibilities presented by emerging technology have proved to be the most enticing to the imagination—and the most difficult to successfully predict. Of course, early visions of a postindustrial technological society were strikingly prescient in many respects. Debates about today's emerging technologies echo many of the utopian and dystopian motifs of twentieth-century science fiction: fears *and* hopes of a "brave new world" of bioengineered humans constructed by exquisite design rather than evolutionary chance; of humans working side by side with intelligent robotic caregivers, surgeons, and soldiers; of digitally enabled "Big Brothers" recording and analyzing our every act; and of the rise of a globally networked hive mind in the "cloud" that radically transforms the nature of human communication, productivity, creativity, and sociality.

Still, we cannot help but smile wistfully at the lacunae of even our most far-seeing science fiction visionaries. In the classic Ray Bradbury tale "The Veldt" (1950), we encounter the existential and moral dilemma of the Hadley family, whose complete surrender to the technological comforts of the "Happy-life Home" has stripped their lives of labor, but also of joy, purpose, and filial love.

In a present marked by the increasingly sophisticated design of "smart homes," Bradbury's story resonates still. It may have taken a few decades longer than he expected, but affluent modern families can now, just like the Hadleys, enjoy a home that anticipates their every personal preference for lighting, room temperature, music, and a perfectly brewed cup of coffee—and the "smart homes" of the future will even more closely approximate Bradbury's vision. We also recognize all too well the Hadleys' parental anxiety and regret when their children, irretrievably spoiled by the virtual world of their interactive playroom, fly into an incandescent rage at the thought of having their electronic amusements removed.

Yet today we can only laugh or cry when Lydia, the children's mother, complains that her surrender to domestic technology has left her without "enough to do" and too much "time to think." No technologically savvy twenty-first-century parent can identify with Lydia Hadley's existential plight.[3] Rather, the promised land of unlimited technological leisure gave way to a reality of electronic overstimulation and hypersaturation, a twenty-four-hour news cycle, and smartphones on which your boss texts you from the eighteenth hole in Dubai while you sit at the dinner table wolfing down takeout, supervising your child's web research on whale sharks, feverishly trying to get caught up on your email as well as your Facebook invitations and your Twitter DMs, and updating the spreadsheet figures your colleagues need for their afternoon presentation in Seoul. For many, the dramatic rise of technology-enabled remote working during the COVID-19 pandemic only amplified digital labor's assimilation of our experience of domestic and personal time. Leisure is one thing our age does *not* afford most modern technology consumers, who struggle each night to ignore the incoming status updates on their bedside devices so that they may grab a few precious hours of sleep before rejoining that electronic day that knows neither dusk nor dawn.

Indeed, the contemporary human situation is far more complicated, dynamic, and unstable than any of the worlds depicted in our first imaginings of a high-tech future. Today, exponential leaps in technological prowess and productivity are coeval with widespread economic stagnation and inequality, terrestrial resource depletion, global pandemics, and rising ecological instability. A global information society enabled by a massive electronic communications network of unprecedented bandwidth and computing power has indeed emerged; but, far from enabling a "new world order" of a utopian *or* dystopian sort, the information age heralds an increasingly *dis*ordered geopolitics and widening fractures in the public commons. The rapid amplification of consumerism by converging innovations and ever-shorter product marketing cycles continues apace; yet, far from ensuring the oft-predicted rise of technocratic states ruled by scientific experts, the relationship between science, governance, and public trust is increasingly contentious and unsettled, with disinformation spreading like wild-fire even as new modes of discovery and truth-seeking are created.[4]

Paradoxically, such tensions appear to be greatest where scientific and technical power has been most successfully consolidated and embedded into our way of life. Consider that the nation that gave birth to Apple, Microsoft, Google, Intel, Amazon, and other tech behemoths has slashed federal funding for basic science research, struggled with declining scientific literacy and technical competence among its population, and adopted increasingly ambivalent, shortsighted, and politicized science and technology policy. Nowhere have the consequences been more devastating than in the realm of public health. In the United States and the United Kingdom, resistance by key government leaders *and* misinformed citizens to scientific advice, paired with unfounded skepticism about technologies from vaccines to masks, has cost countless lives. Contradicting the initial rhetoric that "we are all in this together," a disproportionate share of those losses has been borne by already marginalized populations—ethnic minorities, elderly and disabled persons, those with underlying health conditions, and the working poor. This *despite* the prior existence of detailed "playbooks" designed by leading public health experts to equip government leaders with every available scientific and technological advantage to effectively control a pandemic. No one expected that the playbooks would simply be ignored.

Such complexities remind us that predicting the general shape of tomorrow's innovations is not, in fact, our biggest challenge: far harder, and more significant, is the job of figuring out what we will *do* with these technologies once we have them, and what they will do with *us*. This cannot be done without attending to a host of interrelated political, cultural, economic, environmental, and historical factors that co-direct human innovation and practice. Indeed, a futurist's true aim is to envision not the technological future but our techno*social* future—a future defined not by which gadgets we invent but by how our evolving technological powers become embedded in co-evolving social practices, values, and institutions. Yet by this standard, our present condition seems not only to defy confident predictions about where we are heading but even to defy the construction of a coherent narrative about where exactly we *are*. Has the short history of digital culture been one of overall human progress or decline? On a developmental curve, are we approaching the next dizzying explosion of technosocial innovation, as some believe, or teetering on a precipice awaiting a calamitous fall, as others would have it?[5]

Should it matter whether our future can be envisioned with any degree of confidence? Of course, we might *want* to know where we are and where we are heading, but humans characteristically want a lot of things, and not all of these are necessary or even objectively worthwhile. Could it be that our understandable adolescent curiosity about what awaits us in our century's adulthood is, in the grand scheme of things, unimportant to satisfy? Let us imagine for the sake of argument that given certain efforts, we could better predict the future shape of life in this century. Other than idle curiosity, what reason would we have to make such efforts? Why not just take the future as it comes? Why strain to see any better through the fog of technosocial contingencies presently obscuring our view? There is a simple answer. Our growing technosocial blindness, a condition that I will call *acute technosocial opacity*, makes it increasingly difficult to identify, seek, and secure the ultimate goal of ethics—a life worth choosing, a life lived *well*. Let me explain why.

Ethics, defined broadly as reflective inquiry into the good life, is among the oldest, most universal, and most culturally significant intellectual preoccupations of human beings. Few would deny that humans have always and generally preferred to live well rather than badly, and have sought useful guidance in meeting this desire. Yet the phenomenon of acute technosocial opacity is a serious problem for ethics—and a relatively new one.[6] The founders of the most enduring classical traditions of ethics—Plato, Aristotle, Aquinas, Confucius, the Buddha—had the luxury of assuming that the practical conditions under which they and their cohorts lived would be, if not wholly static, at least relatively stable. While they knew that unprecedented political developments or natural calamities *might* at any time redefine the ethical landscape, the safest bet for a moral sage of premodern times would be that he, his fellows, and their children would confront essentially similar moral opportunities and challenges over the course of their lives.

Without this modest degree of foresight, ethical norms would seem to have little if any power to guide our actions. For even a timeless and universally binding ethical principle presupposes that we can imagine how adopting that principle today is likely to sustain or enrich the quality of our lives tomorrow. Few are moved by an ethical norm or ideal until we have been able to envision its concrete expression in a future form of life that is possible for us, one that we recognize as relevantly similar to, but qualitatively better than, our current one. When our future is opaque, it is harder to envision the specific conditions of life we will face tomorrow that can be improved by following an ethical principle or rule today, and such ideals may fail to motivate us.

While philosophical ethics first emerged in Greece and Asia in the sixth through fourth centuries BCE, the need for ethical guidance as we face our future applies equally to modern systems of ethics. Yet modern ethical frameworks often provide *fewer* resources for mitigating the difficulty posed by an uncertain future than do classical traditions. For example, the ethical framework of eighteenth-century German philosopher Immanuel Kant supplied a single moral principle, known as the *categorical imperative*, which is supposed to be able to resolve any ethical dilemma. It simply asks a person to consider whether she could will the principle upon which she is about to act in her particular case to be universally obeyed by all other persons in relevantly similar cases (Kant [1785] 1997). If she can't will her own "subjective" principle of action to function as a universal rule for everyone to follow, then her act is morally wrong. So if I cannot will a world in which everyone lies whenever it would spare them trouble, then it cannot be right for me to lie.

Though it can be applied to any situation, the rule itself is highly abstract and general. It tells us nothing specific about the shape of moral life in eighteenth-century Europe, nor that of any other time or place. At first, we might think this makes the principle *more* useful to us today, since it is so broad that it can apply to any future scenario we might imagine. Yet this intuition is mistaken. Consider the dutiful Kantian today, who must ask herself whether she can will a future in which *all* our actions are recorded by pervasive surveillance tools, or a future where we *all* share our lives with social robots, or a future in which *all* humans use biomedical technology to radically transform their genes, minds, and bodies. How can any of these possible worlds be envisioned with enough clarity to inform a person's will? To envision a world of pervasive and constant surveillance, you need to know what will be done with the recordings, who might control them, and how they would be accessed or shared. To know whether to will a future full of social robots, you would first need to know what roles such robots would play in our lives and how they might transform human interactions. To will a world where all humans enhance their own bodies with technology, won't you first need to know which parts of ourselves we would enhance, in what ways, and what those changes would do to us in the long run—for example, whether we would end up improving or degrading our own ability to reason ethically? Once even a fraction of the possible paths of technosocial development are considered, the practical uncertainties will swamp the cognitive powers of any Kantian agent, paralyzing her attempt to choose in a rational and universally consistent manner. Kant's categorical imperative has another well-known formulation, which forbids treating rational beings as mere means to our own ends. Though more helpful in certain cases, this too fails to address the technomoral complexities of a world in which, for example, an intelligent, autonomous robot or software agent challenges our concept of a "rational being."

Modern utilitarian ethics of the sort promoted by nineteenth-century British philosophers Jeremy Bentham and John Stuart Mill fares little better by telling us that we may secure the good life simply by choosing from among the available courses of action the one that promises the greatest happiness for all those affected. The problem of discerning *which* course of action promises the greatest overall happiness or the least harm, among all the novel paths of biomedical, mechanical, and computational development open to us, is simply incalculable. The technological potentials are too opaque, and too numerous, to assign reliable probabilities of specific outcomes. Moreover, technology often involves effects on humanity created by the aggregate choices of *many* groups and individuals, not just one person. When we factor in the interaction effects between converging technologies, social practices, and institutions, the difficulty becomes intractable.

In their book *Unfit for the Future*, philosophers Ingmar Persson and Julian Savulescu note that the technological and scientific advances of the twentieth century have further destabilized the traditional moral calculus by granting humans an unprecedented power to bring about "Ultimate Harm"—namely, "making worthwhile life *forever* impossible on this planet" (Persson and Savulescu 2012, 46). We might destroy ourselves with a bioengineered virus for which we have no natural defenses. Carbon dioxide, nitrogen, and phosphorus from large-scale industry and agriculture may acidify our oceans and poison our waterways beyond repair. Or we might unleash a global nuclear holocaust, a risk that experts warn is once again on the rise (Lewis et al. 2014). How can existential risks such as this, scenarios that would ruin any future possibility for happiness, possibly be factored into the calculation?

Moreover, emerging technologies such as nanomedicine and geoengineering in theory have the potential to forestall "Ultimate Harm" to humanity *or* to cause it, and not enough is known to reliably calculate the odds of either scenario. Add to this the fact that engineers and scientists are constantly envisioning new and untested avenues of technological development, and the insolubility of the moral calculus becomes even more obvious. Even John Stuart Mill noted that the practicality of utilitarian ethics relies heavily upon our collective inheritance of centuries of accumulated moral wisdom about how to maximize utility in the *known* human environment (Mill [1861] 2001, 23–25). Even on the timescale of our own lives, this environment is increasingly unstable and unpredictable, and it is not clear how much of our accumulated wisdom still applies.

This leads us to ask: given this unprecedented degree of technosocial opacity, how can humans continue to do ethics in any serious and useful way? The question compels an answer; to abandon the philosophical project of ethics in the face of these conditions would not only amplify the risk of "Ultimate Harm" but violate a deep-seated human impulse. Consider once again Ray Bradbury, whose stories are still among the most widely read and appreciated in the tradition of science fiction. What drives the imagination of a storyteller such as Bradbury, and what makes his stories resonate with so many? Reading his most lauded works, *Fahrenheit 451*, *The Martian Chronicles*, and the collection *The Illustrated Man* (which leads with "The Veldt"), one notices how closely Bradbury's vision tracked human beings of a future Earth, or human descendants of Earth. Why this anthropological fidelity in a writer hardly wanting for imaginative horsepower?

Even the Martians in Bradbury's stories serve as literary foils who expose and reflect upon the distinctive powers, obsessions, and weaknesses of human beings. And why is the human future usually envisioned on a timescale of fifty years, or a hundred and fifty? Why not a thousand years, or ten thousand? Why do so many of Bradbury's tales have a patently ethical arc, driven less by saintly heroes and diabolical villains than by ordinary, flawed humans working out for themselves how well or how poorly their lives have gone in an era defined by rockets, robots, and "televisors"? Here is one plausible answer: Bradbury seemed compelled to imagine how human beings *more or less like himself*, and those he cared about, would fare in the not so distant technological future—to envision the possibilities for *us* living well with emerging technologies, and more often the possibilities for our failing to live well.

All of this is meant to suggest that the ethical dilemmas we face as twenty-first-century humans are not "business as usual," and require a novel approach. Now, it is a common habit of many academics to roll their eyes at the first hint of a suggestion that the human situation has entered some radically new phase. As a prophylactic against overwrought claims of this kind, these sober-minded individuals keep on hand an emergency intellectual toolkit (which perhaps should be labeled "Break Glass in Case of Moral Panic") from which they can readily draw a litany of examples of any given assertion of transformative social change being trumpeted just as loudly a century ago, or five, or ten. This impulse is often well motivated: libraries worldwide are stocked with dusty treatises by those who, from either a lack of historical perspective or an intemperate desire to sell books, falsely asserted some massive seismic shift in human history that supposedly warranted great cultural alarm.

Yet sometimes things really *do* change in ways that we would be remiss to ignore, and which demand that we loosen up our scripted cultural patterns of response. At risk of inviting the scorn of the keepers of academic dispassion, I suggest that this is one of those times. The technologies that have emerged in the last half century have led to the unprecedented economic and physical interdependence of nations and peoples and an equally unprecedented transmissibility of information, norms, ideas, and values. A great many intellectual and cultural scripts are being rewritten as a result—scripts about modern state power, socioeconomic development, labor and human progress, and our relationship with our environment, to offer just a few examples. It follows that the conventional scripts of philosophical ethics must be rewritten as well. While an irreducible plurality of ethical narratives is both inevitable and desirable in a world as culturally diverse as ours, we need a common framework in which these narratives can be situated if humans are going to be able to address these emerging problems of *collective* technosocial action wisely and well. This framework must facilitate not only a shared moral dialogue but also a global commitment to the cultivation of the specific technomoral habits and virtues required to meet this challenge.

The Technomoral Virtues

Fortunately for us, a tradition already exists in philosophy that can provide such a framework. That tradition is *virtue ethics,* a way of thinking about the good life as achievable through specific moral traits and capacities that humans can actively cultivate in themselves.[7] The rich conceptual resources of the classical virtue traditions of Aristotelian, Confucian, and Buddhist ethics, among others, can help us to construct a contemporary framework of *technomoral virtues* explicitly designed to foster human capacities for flourishing with new technologies. Such a framework can be applied to multiple domains of emerging technology, from AI and robotics to biomedical enhancement technology, that are likely to reshape human existence in the next hundred years, assuming that we are fortunate and prudent enough to make it to the twenty-second century.

No ethical framework can cut through the general constraints of technosocial opacity. The contingencies that obscure a clear vision of even the next few decades of technological and scientific development are simply far too numerous to resolve—in fact, given accelerating changes in our present physical, geopolitical, and cultural environment, these contingencies and their obscuring effects are likely to multiply rather than diminish. What this approach offers is not an ethical *solution* to technosocial opacity but an ethical *strategy* for cultivating the type of moral character that can aid us in coping, and even flourishing, under such challenging conditions.

Virtue ethics offers such a strategy in its fundamental orientation to cultivating *practical wisdom,* a form of moral intelligence that enables the skillful, creative, and adaptive modulation of moral judgment and habit to novel or rapidly changing contexts and circumstances. Practical wisdom is the kind of excellence we find in moral experts, persons whose moral lives are guided by appropriate feeling and intelligence, rather than by mindless habit or rote compulsion to follow fixed moral scripts provided by religious, political, or cultural institutions. As noted by the moral philosopher Kongzi (commonly known by his Latinized name, Confucius), the acts of a virtuous person are made noble not simply by their correct content—though that *will* typically respect important moral conventions—but by the singular and authentic moral *style* in which that person chooses to express their virtue. It is this aesthetic mode of personally expressing a moral convention, rule, or script that embodies and presents one's virtue. The person who enacts fixed moral rules rigidly, without style, feeling, thought, or flexibility, is, on this view, a shallow parody of virtue, what the Confucian tradition refers to derisively as the "village honest man" (Yearley 2002, 256).

Even reliably pro-social habits such as truth-telling and law-following fail to guarantee virtue. For while the virtuous person will certainly have such habits, moral intelligence is required to ensure that these habits do not produce acts that violate the moral sense of the situation—for example, mindless obedience to a lawful but profoundly immoral and indefensible order. Actions issuing from the moral habits of a virtuous person—that is, a person with practical wisdom—are properly attuned to the unique and changing demands of each concrete moral situation. In contrast, a person who is prone to thoughtless and unmodulated action is likely to go wrong as often as not; the rookie police officer who runs directly into a hostage situation is foolish and reckless, not courageous, and may well endanger others in the process. Thus moral virtue presupposes knowledge or understanding. Yet unlike theoretical knowledge, the kind of knowledge required for moral virtue is not satisfied by a grasp of universal principles; instead, it requires recognition of the relevant and operative practical conditions. There are imaginable, if rare, circumstances where running into a hostage situation *would* be the wisest and most courageous option for the rookie, even if most of the time such circumstances demand restraint.

Moral expertise thus entails a kind of knowledge extending well beyond a cognitive grasp of rules and principles to include emotional intelligence; keen awareness of the motivations, feelings, beliefs, and desires of others; a sensitivity to the morally salient features of particular situations; and a creative knack for devising appropriate practical responses to those situations, especially where they involve novel or dynamically unstable circumstances. For example, the famous "doctrine of the mean" embedded in classical virtue theories entails that the morally wise agent has a quasi-perceptual ability to see how an emergent moral situation requires a spontaneous and often unprecedented realignment of conventional moral behaviors. Even if it is, as a rule, morally wrong to touch naked strangers without their consent, I had better not hesitate to drag my neighbor's naked, unconscious body out of a burning bedroom when I can do so safely. There is simply no way to capture the full content of all such tacit and embodied moral knowledge in propositional statements or explicit and fixed decision procedures (Nussbaum 1990, 73–74). As Aristotle took pains to note in the *Nicomachean Ethics*, matters of practical ethics by nature "exhibit much variety and fluctuation," requiring a distinctive kind of reasoning that displays an understanding of changing particulars as well as fixed universals (1984, 1094b16). On Aristotle's view, while it is true that rational principles are part of ethics, it is the virtue of practical wisdom that establishes the correct moral rule or principle in any case, rather than wisdom being defined by its correspondence with a prior principle.[8]

Now we are in a position to understand why, if our aim is to learn how to live well with emerging technologies, a virtue ethics approach will generally be more useful than one that relies upon fixed consequentialist or deontological principles. If the practical conditions of ethical life in the fourth century BCE already displayed too much variety and flux for us to rely upon a principle-based ethics, requiring instead an account that articulates the specific virtues of persons who judge wisely and well under dynamic conditions, then the practical uncertainties and cultural instabilities produced by emerging technologies of the twenty-first century would seem to make the contemporary case for virtue ethics that much stronger.[9]

A key phenomenon accelerating the acute technosocial opacity that defines our age is that of *technological convergence*: discrete technologies merging synergistically in ways that greatly magnify their scope and power to alter lives and institutions, while also amplifying the complexity and unpredictability of technosocial change. The technologies most commonly identified as convergent are the fields of applied technoscience referred to as NBIC technologies: nanotechnology, biotechnology, information technology, and cognitive science.[10] Consider just briefly the impact of their convergence on the emerging markets for brain implants, cybernetic prosthetics, replacement organs, lab-grown meat, "smart" drugs, "lie-detecting" or "mind-reading" brain scanners, and artificially intelligent robots—and the panoply of new ethical dilemmas already being generated by these innovations. Now ask which practical strategy is more likely to serve humans best in dealing with these unprecedented moral questions: a stronger commitment to adhere strictly to fixed rules and moral principles (whether Kantian or utilitarian)? Or stronger and more widely cultivated habits of moral virtue, guided by excellence in practical and context-adaptive moral reasoning? I hope I have given the reader cause to entertain the latter conclusion.

We can already enhance the plausibility of this claim by noticing an emerging asymmetry between the moral dilemmas presented by today's converging technologies and the topics that still dominate most applied ethics courses and their textbooks. These textbooks have sections devoted to weighing the ethics of abortion, capital punishment, torture, eating meat, and so on. In each case, it seems reasonable to frame the relevant moral question as "Is x [where x is an act or practice from the preceding list] right or wrong?" Of course, these questions may or may not have definitive answers, and to get an answer one may need to specify the conditions under which the act is being considered—for example, whether alternative sources of nutrition are available to the meat-eater. Still, such questions make sense, and we can see how applying various moral principles might lead a person to concrete answers. Compare this with the question "Is Twitter right or wrong?" or "Is social robotics right or wrong?" There is something plainly odd and ill-formed about such questions, and it is not clear how traditional moral principles could be of any help in answering them.

At this point, the reader will likely object that we are asking about the rightness or wrongness of technologies rather than of acts, and that this is the primary source of our confusion. But notice that it does not actually help things to reform our questions in action terms, such as "Is tweeting wrong?" or "Is it wrong to develop a social robot?" The asymmetry is of a different nature. It is not even that one set of problems involves technology and another does not; after all, technology is heavily implicated in modern practices of abortion and capital punishment. The problem is that emerging technologies such as social networking software, social robotics, global surveillance networks, and biomedical human enhancement are not yet sufficiently developed to be assignable to specific practices with clear consequences for definite stakeholders. They present open developmental possibilities for human culture as a whole, rather than fixed options from which to choose. The kind of deliberation they require, then, is entirely different from the kind of deliberation involved in the former set of problems.

Of course, the line is not a bright one, as emerging technologies also impact long-standing practices where fixed moral principles retain considerable normative force, such as data privacy and copyright protection. Yet it remains the case that very often, the answers for which questions about emerging technology beg are simply not of the yes/no or right/wrong sort. Instead, they are questions of this sort: "How might interacting with social robots help, hurt, or change us?"; "What can tweeting do to, or for, our capacities to enjoy and benefit from information and discourse?"; "What would count as a 'better,' 'enhanced' human being?" It should be clear to the reader by now that these questions invite answers that address the nature of human flourishing, character, and excellence—*precisely the subject matter of virtue ethics.*

What this approach requires, then, is a fuller profile of the technomoral virtues called for in twenty-first-century life.[11] These will not be radically new traits of character, for they must be consistent with the basic moral psychology of our species. Rather, the technomoral virtues are new alignments of our existing moral capacities for justice, courage, honesty, and self-control, along with other virtues adapted to a rapidly changing environment that increasingly calls for collective moral wisdom on a global scale. In these challenging circumstances, the technomoral virtues offer the philosophical equivalent of a blind man's cane. While we face a future that remains cloaked in a technosocial fog, this need not mean that we go into it unprepared or ill-equipped, especially when it comes to matters of ethical life. The technomoral virtues, cultivated through the enduring practices and habits of moral self-cultivation that we can relearn and adapt from the classical virtue traditions, are humanity's best chance to cope and even thrive in the midst of the great uncertainties and vicissitudes of technosocial life that lie ahead.[12] This hope will only be realized, however, if these virtues are more consciously cultivated in our families, schools, and communities, supported and actively encouraged by our local and global institutions, and exercised not only individually but *together*, in acts of collective human wisdom. This is a tall order, but it is not beyond our capabilities.

There is, however, what philosophers call a "bootstrapping problem." Our hope of flourishing in this and coming centuries—or even of securing our continued existence in the face of species-level threats created by our present lack of technomoral wisdom—requires us to act very soon to commit significant educational and cultural resources to the local *and* global cultivation of such wisdom. The approach proposed here, which seeks to draw strength from a diverse cultural pool of historical sources of moral wisdom, can help us accomplish just that. Yet our existing technomoral vices, along with the normal human range of cognitive biases, impede many of us from grasping the depth, scope, or immediacy of the threats to human flourishing now confronting us. Even among those who recognize the dangers, many fail to grasp that the solution must be an *ethical* one. We cannot lift ourselves out of the hole we are in simply by creating more and newer technologies, so long as these continue to be designed, marketed, distributed, and used by humans every bit as deficient in technomoral wisdom as the generations that used *their* vast new technological powers to dig the hole in the first place!

While the first step out of the hole requires reallocating individual, local, and global resources to technomoral education and practice, we can and must make wise and creative use of technology to aid in the effort. Each of the emerging technologies discussed in this chapter has the potential to be designed and used in ways that reinforce, rather than impede, our efforts to become wiser and more virtuous technological citizens. Thus our way out of the hole is a *recursive* procedure, in which traditional philosophical and educational techniques for cultivating virtue are used to generate the motivation to design and adopt new technological practices that shape our moral habits in more constructive ways. These in turn can reinforce our efforts of moral self-cultivation, forming a virtuous circle that makes us even more ethically discerning in technical contexts as a result of increasing moral practice in those domains. This growing moral expertise in technosocial contexts can enable the development of still better, more ethical, and more sustainable technologies. Used as alternating and mutually reinforcing handholds, this interweaving of moral and technological expertise is a practical and powerful strategy for cultivating technomoral selves: human beings with the virtues needed to flourish together in the twenty-first century and beyond.

Motivations of the Approach

My path to thinking about the ethical challenge of emerging technologies in terms of technomoral virtues was steered by a growing unease with unusually rapid changes I observed in my own moral and intellectual habits of attention and self-control, changes that began in the early 2000s as I immersed myself in the then-new digital environments of social media and the mobile app economy. When I first voiced these concerns in 2006 to a class I was teaching on science, technology, and society, my students surprised me with their response. Far from dismissing my worries as silly technophobia, my students responded with overwhelming gratitude, even desperation, for a chance to talk openly about how their own happiness, health, security, and moral character were being shaped by their new technological habits in ways that often bypassed their understanding or conscious choice.

These concerns will be familiar to readers of popular writing on digital culture. Nicholas Carr, Evgeny Morozov, and Jaron Lanier are just a few of the prominent cultural critics who first expressed alarm at the possibility, even likelihood, that our mediatized digital culture could undermine core human values, capacities, and virtues. Carr's *The Shallows* (2010) alerted us to growing evidence of the deleterious cognitive and moral effects that our new digital consumption habits may be having on our brains. Morozov's *The Net Delusion* (2011) and *To Save Everything, Click Here* (2013) powerfully challenged our unreflective faith in technocratic "solutionism." From Lanier, a computer scientist and pioneering innovator of virtual reality technology, came the widely read humanistic manifesto *You Are Not a Gadget* (2010), which lamented the domination of contemporary technosocial life by the increasingly libertarian and anti-humanistic values celebrated by many Silicon Valley technologists: free-market capitalism, consumerism, and reductive efficiency.

Though my views do not align with theirs in every respect, my approach shares with these critics a deeply humanistic and explicitly moralized conception of value. It assumes that the "good life," by which we mean a human future worth seeking, choosing, building, and enjoying, must be a life lived *by* and *with* persons who have cultivated some degree of ethical character. It assumes that this is the *only* kind of human life that is truly worth choosing, despite the perpetual challenges we encounter in building and sustaining such lives. It also holds that a good and choiceworthy life has never been attained in any great measure by isolated individuals, but only by persons who were fortunate enough to enjoy some degree of care, cooperation, and support from other humans, and who were highly motivated to give the same. My approach is therefore fundamentally inconsistent with anti-humanistic and libertarian philosophies, and—if Lanier is right—inconsistent with the philosophy of many of those driving the emerging technological developments it proposes to examine.

Yet my approach carries a resolute hope for the future of human flourishing *with*, not without or in spite of, the technosocial innovations that will continue to shape and enrich our lives for as long as human culture endures. As a scholar who chose out of all possible specialties the philosophy of science and technology, who as a young girl wrote adventure games in BASIC for her Commodore PET and eschewed the Barbie Dream'Vette in favor of Star Wars AT-AT and X-Wing toys, it is simply impossible for me to be anti-technology, personally or philosophically. Indeed, to be anti-technology is in some sense to be anti-human, for we are what we do, and humans have always engineered our worlds as mirrors of our distinctive needs, desires, values, and beliefs. Of course, we are not alone—increasingly, researchers find other intelligent animals such as birds, elephants, and cephalopods reshaping their environments and practices in familiar ways. Perhaps to be anti-technology is also to be anti-life, or anti-sentience. But however widely we share this part of ourselves with other creatures, humanity without technology is not a desirable proposition—it is not even a meaningful one. The only meaningful questions are: which technologies shall we create, with what knowledge and designs, affording what, with whom, for whose benefit, and to what greater ends? These are the larger questions that drive my work. Yet humans lacking the technomoral habits and virtues I seek to describe in my writings could, I think, never hope to answer them. Let us not surrender that hope.

Notes

* This chapter has been adapted from the introduction and chapter 1 of *Technology and the Virtues: A Philosophical Guide to a Future Worth Wanting* (Vallor 2016). I am deeply grateful to the Markkula Center for Applied Ethics at Santa Clara University and to the Arnold L. and Lois S. Graves Foundation, who provided financial support for early phases of this work, and to Lucy Randall at Oxford University Press for her invaluable editorial assistance. Though I should thank many more, I am deeply indebted to Charles Ess, Don Howard, Patrick Lin, Evan Selinger, John Sullins, and P. J. Ivanhoe for their contributions to my thinking for this project.

1. Much contemporary philosophy of technology reflects this view; see Verbeek 2011 for one influential account of how technologies and human social practices are co-constitutive. In this chapter the adjectives *technosocial* and *technomoral* reflect this perspective.
2. The reference here is to the widespread emissions software fraud by Volkswagen publicly exposed in September 2015.
3. The ubiquity of the opposite complaint is highlighted by Turkle (2011, 167).

4. Postindustrial visions of a technocratic future are found in the works of Henri Saint-Simon, Auguste Comte, Thorstein Veblen, and John Dewey, to name just a few.

5. Ray Kurzweil (2005) is the radical standard-bearer for the former, techno-optimist camp of futurists; Bill Joy (2000) is often cited as the contemporary voice of techno-pessimism. Garreau (2005) gives a thorough account of these competing visions, along with several alternatives.

6. See Allenby and Sarewitz 2011 for a related claim.

7. Though I reject Martha Nussbaum's view that the label of "virtue ethics" is so broad as to be vacuous, I take her advice that we who wish to make use of the rich practical resources of virtue traditions not get wrapped up in the art of defining our unique kind, and get on with the business of "figuring out what we ourselves want to say" (1999, 163).

8. See Aristotle 1984 (bk. VI, chap. 13, esp. 1144b20–30). See also MacIntyre 1984, 150–52. This commitment to the dependent status of rational principles of ethics distinguishes Aristotle's view from those who grant the virtues an essential place in ethics, but regard their practical content as derivable from fixed ethical principles; see, for example, O'Neill 1996.

9. Indeed, in their book *The Techno-Human Condition*, Braden Allenby and Daniel Sarewitz (2011) argue that consequentialist and deontological ethics are crippled by the unprecedented and irremediable ignorance of the future that marks this condition—what I have termed *acute technosocial opacity*. Despite their correct diagnosis, Allenby and Sarewitz fail to explicitly acknowledge that the solution they call for is, in fact, not a "re-invented Enlightenment" ethic (187) but a technosocial virtue ethic heavily indebted to classical conceptions of practical wisdom.

10. See Bainbridge and Roco 2006; Khushf 2007; Nordmann 2005.

11. Mine, then, is an explicitly *pluralistic* account—one that willingly sacrifices some theoretical unity in exchange for practical cash value in addressing collective problems of technomoral wisdom. It is also a human-centered account, insofar as I take humans to be the only agents presently capable of deliberating together about the good life; yet an ability to attend and respond to *all* forms of moral worth, including that of nonhumans and the environment, is explicitly identified with the technomoral virtues of empathy, care, and perspective.

12. These virtues do not function *exclusively* for this purpose; individually, they each serve other ethical functions in the various cultural traditions for which they are meaningful. Yet when cultivated *together* they may also serve a global function in helping us to address a global human problem of twenty-first-century life—namely, how to flourish in a condition of acute technosocial opacity and increasing existential risk.

References

Allenby, Brad, and Daniel Sarewitz. 2011. *The Techno-Human Condition*. Cambridge, MA: MIT Press.

Aristotle. 1984. *Nicomachean Ethics*. In *The Complete Works of Aristotle: Revised Oxford Translation*, edited by Jonathan Barnes, 2:1729–867. Princeton: Princeton University Press.

Bainbridge, William Sims, and Mihail C. Roco, eds. 2006. *Managing Nano-Bio-Info-Cogno Innovations: Converging Technologies in Society*. Dordrecht: Springer.

Bradbury, Ray. 1951. "The Veldt." In *The Illustrated Man*, 9–27. New York: Simon and Schuster.

Carr, Nicolas. 2010. *The Shallows: What the Internet Is Doing to Our Brains*. New York: W.W. Norton.

Cellan-Jones, Rory. 2014. "Stephen Hawking Warns Artificial Intelligence Could End Mankind." BBC, December 2 2014, https://www.bbc.co.uk/news/technology-30290540. Accessed April 11, 2021.

de Waal, Frans. 2007. *Chimpanzee Politics: Power and Sex Among Apes*. Baltimore: Johns Hopkins University Press.

Garreau, Joel. 2005. *Radical Evolution: The Promise and Peril of Enhancing Our Minds, Our Bodies—and What It Means to Be Human*. New York: Doubleday.

Hawking, Stephen, Stuart Russell, Max Tegmark, and Frank Wilczek. 2014. "Transcendence Looks at the Implications of Artificial Intelligence—But Are We Taking AI Seriously Enough?" *The Independent*, May 22. http://www.independent.co.uk/news/science/stephen-hawking-transcendence-looks-at-the-implications-of-artificial-intelligence-but-are-we-taking-9313474.html.

Joy, Bill. 2000. "Why the Future Doesn't Need Us." *Wired* 8, no. 4. www.wired.com/wired/archive/8.04/joy.html.

Kant, Immanuel. (1785) 1997. *Groundwork of the Metaphysics of Morals*. Translated by Mary Gregor. Cambridge: Cambridge University Press.

Khushf, George. 2007. "The Ethics of NBIC Convergence." *Journal of Medicine and Philosophy: A Forum for Bioethics and Philosophy of Medicine* 32, no. 3: 185–96.

Kurzweil, Ray. 2005. *The Singularity Is Near: When Humans Transcend Biology*. New York: Penguin.

Lanier, Jaron. 2010. *You Are Not a Gadget: A Manifesto*. New York: Knopf.

Lewis, Patricia, Heather Williams, Susan Aghlani, and Benoit Pelopidas. 2014. *Too Close for Comfort: Cases of Near Nuclear Use and Options for Policy*. Chatham House Report of the Royal Institute of International Affairs. London: Chatham House.

MacIntyre, Alasdair. 1984. *After Virtue*. 2nd ed. Notre Dame, IN: University of Notre Dame Press.

Mill, John Stuart. (1861) 2001. *Utilitarianism*. Indianapolis, IN: Hackett.

Morozov, Evgeny. 2011. *The Net Delusion: The Dark Side of Internet Freedom*. New York: PublicAffairs/Perseus Books.

Morozov, Evgeny. 2013. *To Save Everything, Click Here: The Folly of Technological Solutionism*. New York: PublicAffairs/Perseus Books.

Nordmann, Alfred. 2005. *Converging Technologies: Shaping the Future of European Societies*. Luxembourg: Publications Office of the European Union.

Nussbaum, Martha C. 1990. *Love's Knowledge*. Oxford: Oxford University Press.

Nussbaum, Martha C. 1999. "Virtue Ethics: A Misleading Category?" *Journal of Ethics* 3: 163–201.

O'Neill, Onora. 1996. *Towards Justice and Virtue: A Constructive Account of Practical Reasoning*. Cambridge: Cambridge University Press.

Persson, Ingmar, and Julian Savulescu. 2012. *Unfit for the Future: The Need for Moral Enhancement*. Oxford: Oxford University Press.
Turkle, Sherry. 2011. *Alone Together: Why We Expect More from Technology and Less from Each Other*. New York: Basic Books.
Vallor, Shannon. 2016. *Technology and the Virtues: A Philosophical Guide to a Future Worth Wanting*. New York: Oxford University Press.
Verbeek, Peter-Paul. 2011. *Moralizing Technology: Understanding and Designing the Morality of Things*. Chicago: University of Chicago Press.
Yearley, Lee H. 2002. "An Existentialist Reading of Book Four of the *Analects*." In *Confucius and the Analects: New Essays*, edited by Bryan Van Norden, 237–74. New York: Oxford University Press.

5

Mindful Technology

Mike W. Martin

Mindfulness has become a popular virtue. No longer just a fancy word for attentiveness, *mindfulness* denotes a wide-ranging excellence that promotes stress relief, emotional control, rational decision-making, concentration at work and at school and in sports, and—my interest—skills in developing and using technology. Although Buddhists have long celebrated mindfulness, recent health psychologists sing fuller-throated paeans. One therapist declares that "mindfulness frees us to act more wisely and skillfully in our everyday decisions" and provides "the solution" to countless daily difficulties (Siegel 2010, 34). Another prominent psychologist traces most problems to an absence of mindfulness: "Virtually all of our problems—personal, interpersonal, professional, and societal—either directly or indirectly stem from mindlessness" (Langer 2014, xiii). Such claims are overblown, but I agree that mindfulness warrants attention in thinking about technology.

Mindfulness and *mindful technology* have myriad meanings. I clarify some of them and offer a framework for understanding others. As a working definition, mindfulness is paying attention to what matters in light of relevant values. When those values are sound and applied intelligently, mindfulness becomes a virtue. As this definition implies, mindfulness is not a stand-alone virtue. Instead, its meaning and substance depend on how it implements additional values that specify what matters in particular situations or in general. Insofar as those values are contested, the full meaning of mindfulness is itself subject to disagreement. In this way, mindfulness is a secondary and dependent virtue, and it is nothing like a panacea.

After locating mindfulness within virtue ethics, I distinguish four concepts of mindfulness that recur in discussions of technology. As sample applications, I discuss the use of personal digital technologies; creativity and innovation in designing technology; responsible engineering; and citizenship in a technological society. Along the way I connect mindfulness to my previous writings on professionalism, engineering ethics, creativity, and mental health. Throughout I understand technology broadly to include devices (e.g., machines and tools), knowledge about making and operating devices and technical systems, and

Mike W. Martin, *Mindful Technology* In: *Science, Technology, and Virtues*. Edited by: Emanuele Ratti and Thomas A. Stapleford, Oxford University Press. © Oxford University Press 2021. DOI: 10.1093/oso/9780190081713.003.0006 practices and organizations where devices and techniques are used (Tiles and Oberdiek 1995, 5–9).

Mindfulness in Virtue Ethics

To be mindful is to be attentive, alert, heedful, and vigilant; it is to manifest discriminating perception and awareness; it is to remember, bear in mind, and conscientiously implement important truths and values. These everyday meanings provide wide latitude for psychologists and spiritual advisors to develop more specialized concepts. Is there any core or essential meaning? Perhaps not, but for my purposes one theme stands out: mindfulness is paying attention to what matters in light of relevant values (Martin 2020). To highlight the centrality of values, I call this working definition *value-based mindfulness*. Even by itself, *paying attention* alludes to values that indicate what is significant or interesting, in contrast with our default *wandering mind*, which flits among constant distractions (Goleman 2013, 39). Robert Nozick believes that "the fundamental evaluative activity is selectivity of focus, focusing here rather than there" (Nozick 1989, 120). Certainly *mindful* attention involves evaluation, as Eric Harrison suggests: "When we become mindful of something, we automatically evaluate it: How much longer shall I stay with this? How much attention does this deserve?" (Harrison 2017, 125). In any case, value-based mindfulness provides a conceptual hub to which myriad definitions of mindfulness attach.

The values that indicate what matters in general and in specific situations can be specified from alternative perspectives. In formal terms, two elemental reference points are pertinent: (1) an individual's (or group's) de facto values, and (2) values that are permissible or justified in some way. When the reference point is an individual's values, I speak of *personal-mindfulness*: paying attention to what matters to individuals (or groups) in light of their values. When the reference point is a set of justified or permissible values I speak of *virtue-mindfulness* (mindfulness as a virtue): paying attention to what matters in light of values that are sound, at least in the minimal sense of permissible—not immoral, not irrational, and not unhealthy.

Paying attention to what matters need not involve observing or contemplating the values that make it matter. To be sure, in some contexts these values are implied, as in mindful attention to a work of art (André 2011), mindful deliberations about a moral issue, or mindful self-scrutiny while undergoing cognitive behavioral therapy. But more often, paying attention implies *being attuned to* what matters by manifesting appropriate caring, perception, awareness, remembering, reasoning, emotion, attitude, and conduct. Thus, mindful drivers are attuned in these ways to speed, stop signs, other vehicles, pedestrians, and additional aspects of driving safely. Mindful engineers pay attention to a host of considerations of safety, law, quality, costs, intended use of products, and unintended side effects. And mindful engagement with technology consists in attunement to what is important when using, maintaining, purchasing, or participating in developing technology.

In general, virtue-mindfulness deserves to be included in discussions of virtue ethics, which is itself an ambiguous expression. In a narrow sense, *virtue ethics* refers to virtue-theoretic theories—that is, ethical theories in which good character is morally foundational, rather than rules, right actions, and rights. In a wider sense, *virtue ethics* refers to (1) all studies of good character, regardless of whether they regard good character as morally foundational, and (2) value perspectives emerging from or guiding those studies. I intend the wider sense. As a pragmatist and practical ethicist, I am interested in how virtue-mindfulness functions in any defensible moral outlook and any complete ethical theory, which typically will include some account of good character as well as accounts of right conduct, human rights, and desirable communities and societies (Dewey 1960, 15).

Although I employ the wider sense of virtue ethics, I appreciate that virtue-theoretic ethicists ignited virtue ethics during the late twentieth century. These ethicists sought a fresh ethical framework to counter moral fragmentation and to surmount the endless moral disagreements over moral rights and rules that have been raging since the Enlightenment (e.g., MacIntyre 2007). It immediately became clear, however, that virtue-theoretic ethics takes many forms, and there is no consensus about which is best. Certainly there are hundreds of virtues, each of which can be understood in different ways, as well as distorted by prisms of power, class, gender, and race. Moreover, virtues are frequently in tension with each other—for example, honesty with kindness, justice with mercy, equality with freedom, and courage with humility. Typically, attempts are made to resolve these tensions by establishing a few virtues as cardinal, but these attempts only generate deeper disagreements about which virtues are cardinal in light of competing ideals of good lives. For example, the classical Greeks celebrated wisdom, courage, temperance, and justice, with wisdom supreme, but Christians exalt faith, hope, and love, with love primary.

Furthermore, any hierarchy of virtues can be upset by new social challenges, including those caused by disruptive technologies. For example, as we transition to fully automated cars in order to prevent the hundreds of thousands of deaths worldwide that are caused each year by poor drivers, should we adopt an altruistic algorithm that saves the pedestrian at the cost of the driver (in tragic situations where one or the other must die) or a self-interested algorithm that saves the driver? Even the new celebration of mindfulness can be understood as a response to lives that are increasingly frenetic, fragmented, and deluged with distractions. In addition, new intellectual influences reshape how the virtues are understood, especially influences from psychology, psychiatry, and other health sciences during the past century (Rieff 1987; Martin 2006). Of particular interest, "positive psychologists" now study the virtues openly and extensively (e.g., Peterson and Seligman 2004), including mindfulness.

Two additional complications arise when fitting virtue-mindfulness (or any virtue) into virtue ethics. First, virtues are character traits: features of persons (or groups) shown in habits of conduct, emotion, attitude, and ways of relating to other people. But words for the virtues also apply to individual actions and states of mind. For example, both persons and actions can be honest, courageous, and mindful. This dual usage prompts a question: when virtue words are applied to actions, do they imply that the action manifests the corresponding trait in the agent? For example, in calling an act honest, courageous, or mindful, do we imply that it is produced by an agent who is habitually honest, courageous, or mindful? Aristotle thought so: "Virtuous acts are not done in a just or temperate way merely because *they* have a certain quality, but only if the agent also acts in a certain state, viz. (1) if he knows what he is doing, (2) if he chooses it, and chooses it for its own sake, and (3) if he does it from a fixed and permanent disposition" (Aristotle 1976, II.4, 1105a29–30). The last criterion, acting from a fixed and permanent disposition, is suitable for angels and inspiring for humans. But countless psychological studies demonstrate that virtuous conduct is dramatically influenced by situational factors as much as by fixed dispositions (Doris 2002). Contrary to Aristotle, sometimes it makes sense to depict actions and persons as honest, courageous, or mindful without a "fixed and permanent disposition" being involved.

Second, does virtue always mean *moral* virtue, or are there also non-moral virtues? Specifically, does virtue-mindfulness always amount to *moral-mindfulness*? Aristotle understood virtue (*arete*) widely to include non-moral excellences such as intelligence and athletic skill. He also distinguished moral virtues, such as justice and generosity, from intellectual virtues, such as rationality and creativity. Some virtue ethicists reject his moral-intellectual distinction, perhaps counting his intellectual virtues as moral (e.g., Zagzebski 1996), or perhaps refusing to count intellectual excellences as virtues. Other virtue ethicists are willing to count all kinds of human excellences as virtues, so long as they are not patently immoral, irrational, or unhealthy. That is my view: virtue-mindfulness can be manifested in pursuing any form of excellence based on sound values that exclude immorality, irrationality, and unhealthiness. In this way, virtue-mindfulness can be a moral virtue or an aesthetic, athletic, historical, therapeutic, or religious virtue. Moral and non-moral values interweave in complicated ways, and I will not attempt to draw a sharp line between them. I will understand morality broadly as including moral requirements (such as duties or respect for rights) and ideals and features of character that may influence all areas of lives.

In sum, I understand the *virtue* in virtue-mindfulness broadly. Virtue-mindfulness can refer to a trait and habit: a feature of individuals who tend to pay attention to what matters in light of sound values. It can refer to an act, activity, or state of mind: paying attention to what matters on a particular occasion, in light of relevant sound values. Virtue-mindfulness helps implement all types of values, moral and non-moral, that enter into the kaleidoscope of good lives. Most important, virtue-mindfulness is not a stand-alone virtue. Its meaning and substance depend on additional sound values, including other virtues, that specify what matters. In this way, it is a "dependent virtue" (Slote 1983, 63). Accordingly, any discussion of mindful technology should spell out the relevant values, and those values will usually be subject to some disagreement.

Influential Concepts of Mindfulness

Value-based mindfulness, personal-mindfulness, and virtue-mindfulness are skeletal concepts. They allude to values without fully specifying them. In contrast, more robust concepts of virtue-mindfulness build in specific values and reflect particular value perspectives. I distinguish four robust concepts that recur in discussions of mindful technology. Three of them are developed by psychologists and the fourth by Buddhism, a religion with a psychological bent that has drawn its practitioners into dialogue with psychologists. Buddhist right mindfulness is explicitly tethered to moral and spiritual values. In contrast, psychological concepts of mindfulness typically disguise their connections to values. In addition, psychologists camouflage moral values under expansive concepts such as normality (decency), self-esteem (self-respect), emotional intelligence (responsibility), personal development (moral development), personality (character), and especially mental health—for example, the World Health Organization's definition of health as "complete physical, mental and social well-being" (United Nations 1946), a notion broad enough to include all the virtues as well as happiness and a host of additional values.

Value-Suspending Mindfulness

The most famous psychological concept of mindfulness was developed by Jon Kabat-Zinn in connection with therapeutic meditation. According to Kabat-Zinn, "Mindfulness means paying attention in a particular way: on purpose, in the present moment, and nonjudgmentally" (Kabat-Zinn 2014, 4).

"Nonjudgmentally," he stipulates, implies that all value judgments—of right and wrong, good and bad, likes and dislikes, praise and blame—are set aside during meditation: practicing mindfulness "involves suspending judgment and just watching whatever comes up, including your own judging thoughts, without pursuing them or acting on them in any way" (Kabat-Zinn 2013, 21; see also 2014, 55–56). Hence, I refer to his concept as value-suspending mindfulness. Although his doctoral degree is in molecular biology, Kabat-Zinn's impact on psychology and popular culture is remarkable. Since 1979, when he opened the first clinic to scientifically study mindfulness-based meditation for stress relief, he has inspired hundreds of additional clinics that use meditation to treat an array of ailments. His influence is even reflected in the entry for "mindfulness" that now appears in *Merriam-Webster's Collegiate Dictionary* (2014): "the practice of maintaining a nonjudgmental state of heightened or complete awareness of one's thoughts, emotions, or experiences on a moment-to-moment basis."

Kabat-Zinn associates mindfulness with meditation, but he does not equate them. Mindfulness is a mental activity or state of mind, whereas therapeutic meditation is a practice that employs and cultivates that activity and state. Hence, Kabat-Zinn is not being redundant when he names his therapeutic technique "mindfulness-*based* meditation." Therapeutic meditation takes many forms, the simplest of which is to concentrate on our breathing while slowing and deepening it; when value-charged thoughts and emotions enter consciousness, we briefly take note of them and set them aside in order to return our attention to our breathing. The mindfulness part of the meditation consists of close attention to and careful awareness of what occurs in the moment. Many individuals experience a calming effect produced by a combination of the physiological effects of deep breathing, an enhanced sense of immediate control, and the reduction of anxieties by buffering present experience from both worries about the future and regrets about the past. Some practitioners testify that mindfulness-based meditation increases their self-knowledge (Harari 2018, 314–23).

On the surface, value-suspending mindfulness is the exact opposite of value-based mindfulness. At a deeper level, however, values are always present in the background of both forms. For one thing, the emphasis on paying attention to breathing and immediate experience manifests valuing serenity and self-awareness in the moment. For another thing, Kabat-Zinn emphasizes that the (successful) practice of mindfulness-based meditation finds its rationale and motivation in a "personal vision," a "vision that is truly your own—one that is deep and tenacious and lies close to the core of who you believe yourself to be, what you value in your life, and where you see yourself going" (Kabat-Zinn 2014, 76). Taken by itself, this emphasis on personal values reflects personal-mindfulness. Yet Kabat-Zinn repeatedly slides to virtue-mindfulness in his popular writings, which celebrate health, compassion, tolerance, peace, and love—values that initially attracted him to Buddhist meditation (Gethin 2013, 268). He also recommends "loving-kindness meditation," which involves reflecting on moral and spiritual values, thereby departing from his official definition of mindfulness as nonjudgmental. I believe he fails to see these inconsistencies because he slides among different senses of "nonjudgmental": (1) suspending all value judgments (in his official definition), (2) suspending unduly negative judgments, (3) suspending all negative judgments, and (4) making positive valuations that are compassionate, kind, and loving. In any case, his value-rich writings exemplify how psychological concepts of mindfulness conceal moral values by embedding them in holistic conceptions of health.

Flexibility-Mindfulness

Since the late 1970s, Ellen J. Langer has developed a strikingly different concept of mindfulness in connection with problem-solving and decision-making. According to her concept of flexibility-mindfulness, as I call it, there are three "key qualities of a mindful state of being: (1) creation of new categories; (2) openness to new information; and (3) awareness of more than one perspective" (Langer 2014, 64). I interpret her concept as a version of virtue-mindfulness, albeit an anemic version. Flexibility-mindfulness builds in values of imagination and openness to new ideas and perspectives, but it fails to ground those values in broader values of rationality, responsibility, and intelligence. An imaginative and well-informed sadistic terrorist could be mindful in her sense—think of Osama bin Laden.

In practice, however, Langer invariably presupposes additional background values that intimate a more robust concept of virtue-mindfulness. Consider her depiction of a mindless decision made by the pilot of a jumbo jet (Langer 2014, 5–6). The pilot runs through the preflight checklist in a routine, habitual manner. The flight is scheduled to leave from a Florida location where the weather is usually warm. Acting from habit rather than alertness to the situation, the pilot leaves the plane's anti-icer off and remains oblivious to the ice forming on the wings. The plane crashes after takeoff, killing seventy-four passengers. We can agree that the pilot is not being mindful, but in doing so we are using a more robust concept of mindfulness than Langer offers. A virtue-mindful pilot would pay attention to the ice not only because it qualifies as new information but also because the ice matters in light of sound values—specifically, the pilot's responsibility for the safety of the passengers and plane. Suppose the pilot notices the ice and appreciates the danger it poses but then takes off without turning on the anti-icer. The pilot could be open to relevant information and different viewpoints but utterly callous, reckless, and perhaps murderous. I suspect that Langer fails to connect mindfulness explicitly to moral values because she reduces morality to merely subjective and relative preferences (Langer 2006, 73).

Flow-Mindfulness

Another cluster of concepts, sometimes called task-mindfulness, highlights concentrated attention during activities (beyond meditation and particular decisions). Famously, Mihaly Csikszentmihalyi defines "flow" as immersion in activities in ways that smoothly integrate one's values and skills in pursuing goals. The activities involve steady feedback about progress, enjoyment, a sense of meaning, minimal anxiety, and much self-confidence. Usually there is no explicit self-awareness, although if we periodically assess how we are doing, "the evidence is encouraging: 'You are doing all right.' The positive feedback strengthens the self, and more attention is freed to deal with the outer and inner environment" (Csikszentmihalyi 1990, 39). Most important, flow activities involve careful attention to what matters during flow, and hence I refer to it as flow-mindfulness.

Insofar as it implements one's values, flow-mindfulness is an instance of personal-mindfulness. Yet when the values are involved are sound, or justified in some way, flow-mindfulness also becomes virtue-mindfulness—for example in connection with the professions: "If the surgeon's mind wanders during an operation, the patient's life is in danger. Flow is the result of intense concentration on the present" (Csikszentmihalyi 1996, 112). Moreover, Csikszentmihalyi celebrates flow as "optimal experience," the kind of desirable experience that promotes excellent work, mental health, happiness, creativity, and overall well-being. To that extent, flow-consciousness promotes or becomes virtue-mindfulness. At the same time, Csikszentmihalyi is careful to say that flow is not always morally desirable overall, and that criminals and sadists could be in flow while engaging in harmful activities: "The flow experience, like everything else, is not 'good' in an absolute sense. It is good only in that it has the potential to make life more rich, intense, and meaningful; it is good because it increases the strength and complexity of the self. But whether the consequence of any particular instance of flow is good in a larger sense needs to be discussed and evaluated in terms of more inclusive social criteria" (Csikszentmihalyi 1990, 70).

Right Mindfulness

Religious traditions develop concepts of virtue-mindfulness rooted in their values, beliefs, and practices. The most famous religious concept (or set of concepts) is right mindfulness. Although Buddhism takes different forms that influence how mindfulness is understood, a core concept of right mindfulness is found in the earliest Buddhist texts: clear and discerning comprehension of the spiritual values and truths taught by the Buddha, combined with conscientious action upon them in the moment. As such, mindfulness is a key element in the

Noble Eightfold Path to spiritual enlightenment, and it is intimately connected to values of compassion, tolerance, peace, and nonviolence. (This is true at least in principle, if not always in practice—witness the current intolerance against Muslims in the Buddhist country of Myanmar [Harari 2018, 310].)

I cite Buddhist concepts because of their influence and because they illustrate explicit virtue-mindfulness (in contrast with the more tacit virtue-mindfulness in psychologists' concepts). I emphasize, however, that countless moral thinkers and spiritual traditions have developed concepts of virtue-mindfulness without using the word. Words (linguistic units) other than *mindfulness* are often used to specify concepts (ideas) of mindfulness. Thus, *right mindfulness* is an expression using English words to translate *sati*, as used by the Buddha in the sixth century BCE (Bodhi 2013, 23). Again, Henry David Thoreau's account of "wakefulness" in *Walden* (2004) is regarded by Kabat-Zinn and others as a discussion of mindfulness, even though Thoreau does not use that word. And the Stoics' concept of "moral vigilance" is akin to Buddhist right mindfulness (Hadot 1995, 84). Arguably, concepts of virtue-mindfulness are found in all world religions, frequently linked to meditative practices such as prayer, reading scripture, ritual confession, and contemplating the sacredness of nature.

In sum, value-suspending mindfulness, flexibility-mindfulness, flow-mindfulness, and right mindfulness illustrate how different thinkers and traditions develop different technical concepts for their purposes. The concepts have various similarities but also differences: for example, both flexibility-mindfulness and flow-mindfulness are focused outwardly on activities, whereas value-suspending mindfulness is more focused inwardly on our experience. Directly or indirectly, each concept connects with the core idea of value-based mindfulness as paying attention to what matters in light of relevant values. Nevertheless, the differences among the concepts matter greatly, as do the particular values in question. Accordingly, when these concepts of mindfulness are invoked in discussing technology, two questions should be salient: what is the intended meaning of mindfulness, and which values indicate what we should be mindful of in connection with using and creating technologies? It is best to keep relevant moral values center stage, rather than to shunt them into the background.

Using Personal Digital Devices

Mindfulness is widely discussed in connection with personal digital devices such as cellphones and laptops. The influence of psychology is usually obvious, although different psychological concepts are typically invoked with minimal differentiation. In my view the central question should be: which values ought to guide the use of personal digital technologies, and hence which values should enter into understanding virtue-mindfulness in using the technologies? Yet psychological concepts often disguise the relevant values. Technical efficiency and productivity might be highlighted in connection with mindful work, but more often values of mental health and personal meaning are invoked in ways that might conceal the relevance of moral values. And personal-mindfulness and virtue-mindfulness are frequently conflated.

Consider David M. Levy's engaging book *Mindful Tech*. Levy is a computer scientist who teaches in the Information School of the University of Washington. His book begins with a comic vignette of mindless behavior. Caught on a shopping-mall video recording that went viral, a woman is talking on her cellphone held in one hand and carrying a shopping bag in the other hand. She trips and does a somersault into a fountain, then quickly recovers, steps out of the water, and continues walking. Presumably the woman values her present activities, as well as her safety and esteem. In any case, Levy initially seems to intend personal-mindfulness: paying attention to what matters in light of one's values, whatever those values might be. This suggestion fits most, although not all, of Levy's book. Thus, he sympathetically quotes a characterization of mindfulness offered by one of his students: "When we are mindful we choose to pay attention to what is explicitly important to us" (Levy 2016, 26).

Levy says that mindfulness combines "two forms of attention," which he calls "task focus and self-observation" (Levy 2016, 4). His "task focus," I suggest, is a version of flow-mindfulness. It involves putting our personal values into play in ways that elicit our skills and produce enjoyment and meaning. The main focus is on the task rather than on ourselves, although engaging in the task is typically accompanied by tacit self-awareness. *Self-observation* transforms tacit self-awareness into explicit attention to ourselves, including how we feel and act when using digital devices. Much of Levy's book is organized around "exercises" designed to help us discover, clarify, and improve how we employ digital devices. Kabat-Zinn's value-suspending mindfulness during meditation is one such exercise, although Levy relegates it to an appendix (Levy 2016, 185). Most exercises consist of observing what we are doing while using digital technologies, recording what we observe, reflecting on our observations, formulating personal guidelines for future behavior, and discussing our discoveries with others. Other exercises are experiments in episodically unplugging ("fasting") from technologies for varying lengths of time. (For an opposing stance on unplugging, see Rohan Gunatillake's *Modern Mindfulness* [2017].)

Although personal-mindfulness dominates *Mindful Tech*, virtue-mindfulness is intimated when Levy states his broader aims. He seeks to help us in "using our attention wisely," and to that end we need to "figure out how to make wise choices, and to figure out what constitutes a wise choice, so we can use our digital tools to their best advantage, and to ours" (Levy 2016, 3). Wise choices, he indicates, create balance between the Fast and Slow Worlds in which we live. The Fast World is online existence that tends to be "fast-paced, crazy-busy, information-intensive," often to the point of causing great stress, addiction, and other unhealthy distortions (Levy 2016, ix). The Slow World is slower-paced, calmer, richer in contemplative moments, and centers on meaningful personal relationships, hobbies, sports, religion, politics, and travel. Both worlds are important, but they are often in tension. Wisdom enables us to find a proper balance—that is, a wise configuration of time, energy, and commitment to the two worlds.

In any ordinary sense, wisdom implies understanding important values and truths, and implementing them with good judgment. Levy trades on this ordinary sense, and hence the key question is what values he uses to define wisdom in using digital devices. He specifies two criteria: health and effectiveness, including effectiveness in integrating the Fast and Slow Worlds (Levy 2016, 2). Yet he also says that each of us determines what these criteria imply in practice: "By paying attention to how you use your cell phone, how you handle email, how you feel when you are on Facebook or Pinterest, or when you multitask, you will be able to see which aspects of your current online practices are working well and which aren't" (Levy 2016, 3). This passage seems to return to personal-mindfulness, albeit enriched by self-reflection in light of what we most desire, with no mention of moral values that make demands on us and inspire better lives. Yet invariably, his concepts of health and effectiveness remain sufficiently broad to conceal moral values associated with wisdom.

Levy's type of self-knowledge cannot by itself produce wisdom in the more robust sense of understanding what is good and right, and implementing that understanding with good judgment. Such wisdom requires centering our lives on sound values, not merely being aware of what we are feeling and doing as we pursue whatever we happen to desire. Suppose that individuals with Levy-type self-knowledge deliberately choose to use their mindfully refined digital skills to bully online, engage in hate speech online, steal and violate privacy by hacking into other people's computers, disseminate false political propaganda, sacrifice their families, or place themselves and others at risk (as with the acrobatic woman in the video). No doubt Levy believes that such actions are unwise, but his reduction of wisdom to subjective self-awareness does not provide support for that belief. He largely assumes his readers and students are morally decent, and he conceals moral values under the headings of mental health and effectiveness. Those values should be kept front and center.

In places, Levy gestures toward an explicit recognition of sound values. He cites scientific studies showing that using personal digital devices while driving is distracting and dangerous, adding that on this topic "there really does appear to be a right answer" to how we ought to behave (Levy 2016, 178). Why "appear"?

Why not instead an emphatic condemnation of patently irresponsible behavior that threatens lives, including our own? Levy is as reluctant to make such explicit moral value judgments, as are the psychologists who influence his discussion. Thus he calls for "honest self-reflection" about unhealthy addictions to digital technologies, and he even raises issues of rudeness in using digital technologies during class or a business meeting (Levy 2016, 176–82). Yet repeatedly moral values are concealed or, much worse, shunted aside as mere matters of personal preferences. In general, discussions of the wise uses of digital technologies need to keep alive a vision of moral values as more than mere preferences and personal feelings.

Creating Technology

Mindfulness is often claimed to promote creativity, but in what sense of mindfulness, and what sense of creativity? Creativity, to begin with, is sometimes defined as purposefully generating *new* products (artifacts, ideas, solutions, etc.), where purposefulness allows for serendipity (Weisberg 2006, 60–65). More often, however, it is defined as purposefully generating products that are new *and* valuable—useful, interesting, or desirable in some other way; creative processes are then defined as those that bring about valuable new products. I favor that definition (Martin 2007), and clearly it has a direct tie to value-based mindfulness during the creative process. Values enter in mindfully assessing products and also during the creative process: "The creative person is sensitive to and acts in the light of reasons. It is her responsiveness to reasons that grounds her judgments and actions in recognizing what is new and valuable in the relevant domain" (Kieran 2014, 127). Always there is some reference point in determining what is valuably new, or even merely new. With individual-relative creativity, the product is valuably new relative to an individual's previous ideas, accomplishments, or level of development. With domain-relative creativity, the product is valuably new relative to a domain of activity, such as engineering or science. There is also group-relative creativity and history-relative creativity (Boden 2004). When psychologists celebrate mindfulness as a means to greater creativity, they generally intend individual-relative creativity, albeit with a hint that domain-creativity might ensue.

As for mindfulness, all three psychological concepts might be invoked. First, invoking Kabat-Zinn's value-suspending mindfulness, one writer on creativity says mindfulness is "paying full conscious attention to whatever thoughts, feelings and emotions are flowing through your mind without harsh judgment or criticism" and "being fully aware of whatever is happening in the present moment and not being trapped in the past or worrying about the future" (Penman

2015, 8). I do not doubt that some individuals improve their individual-relative creativity by using meditative practices involving value-suspending mindfulness for relaxation and improving concentration. Some might even improve their domain-creativity. Steve Jobs is often cited in this connection, although he was probably inspired more by Asian religions than by psychology. In general, as Steve Jobs also illustrates, it seems likely that value-laden mindfulness contributes vastly more to domain-relative creativity than value-suspending mindfulness does.

Second, flow and flow-mindfulness contribute to creativity, according to Csikszentmihalyi (1996). When flow-mindfulness is defined by reference to an individual's values, justified or not, it might seem to promote individual-relative creativity, if only by enabling the person to remain enjoyably involved in activities for longer periods of time. Moreover, when the individual's values are those of science or technology (or another domain), flow-mindfulness can contribute to domain-creativity. Immersion in activities with concentration and focus on what matters, using skills that match valuable (and valued) goals, and with lessened self-absorption, is integral to acts of creativity in science and engineering (as are periods of rest and recreation in gestating ideas). In general, studies of creativity report mixtures of intrinsic and extrinsic motivation. Extrinsic motivations are self-oriented in that the preoccupation is seeking and valuing goods external to the creative activity, such as rewards of money and recognition. Intrinsic motivation is desiring and valuing goods internal to the activity and its domain. It is "the motivation to engage in an activity primarily for its own sake, because the individual perceives the activity as interesting, involving, satisfying, or personally challenging; it is marked by a focus on the challenge and the enjoyment of the work itself" (Collins and Amabile 1999, 299). The mindfulness involved in creativity may or may not be based on moral values, but it typically involves values beyond narrow self-absorption. At the same time, non-moral motivations, such as fame and fortune, sometimes contribute to moral contributions (Martin 2007, 39–49; see also Martin 2000, 21–28).

Third, Langer's flexibility-mindfulness, with its emphasis on creating new ideas and being open to new information and viewpoints, is certainly useful in thinking about creativity. Essentially it is a clarion call to free ourselves from rigid mindsets by acquiring relevant information and being open to fresh alternatives. Yet it is crucial to add to her view that the openness serves additional sound values that are implemented with expertise and good judgment in grappling with specific problems—something Langer tacitly does, as we saw in her example of the pilot. When Langer discusses creativity she focuses on process rather than results, especially the process of breaking habitual mindsets. She also celebrates authentic self-expression and downplays socially valued outcomes and solutions, insisting that both values and evaluations of outcomes are merely subjective and relative (Langer 2006, 73). Yet even in art, and certainly in science and technology, the kinds of mindfulness that lead to creativity need to be based on at least some notion of sound values. Regarding science and technological development, creativity is a matter of valuable newness, where the values include advances in understanding, solving significant problems, and at least potentially advancing human welfare. Accordingly, Langer's flexibility-mindfulness needs to be enriched in the direction of virtue-mindfulness.

Langer proclaims, "Everybody has an equal talent for everything" (Langer 2006, 171). In general, psychologists' discussions of creativity, at least in popular books, are infused with such optimism. One origin of this optimism lies in concepts of mindfulness that weaken the connection with sound values. Values carry demands for developing relevant forms of expertise, and for exercising that expertise with good judgment in implementing the values. Psychological concepts of mindfulness are potential aids in fostering creativity when they are framed as versions of value-based mindfulness. They are not, however, esoteric techniques that replace the need for expertise, mastery of disciplines, development of good judgment, and respect for moral and other ideals relevant to particular endeavors.

Responsible Engineering

The virtue-mindfulness of professionals is firmly anchored in responsibilities and ideals that guide their professions. Thus, mindfulness in medicine requires focusing with due care on patient needs while respecting their rights (Epstein 2017); mindfulness in business implies working efficiently toward corporate goals while being sensitive to the needs of customers, colleagues, employees, and others (Gelles 2016, 23); and mindfulness in commercial aviation is rooted in due attention to what is mandated by flight procedures. Likewise, virtue-mindfulness in engineering is understood in light of sound values and virtues that guide engineers in designing, testing, manufacturing, and marketing, monitoring, and maintaining technological products and processes.

Roland Schinzinger and I did not explicitly employ a concept of mindfulness in *Ethics in Engineering* (2005), but much of what we wrote connects with virtue-mindfulness in the development of innovative technologies. We introduced a moral model of engineering as social experimentation. The core idea is that most engineering projects, both routine and creative ones, are carried out without full knowledge. Uncertainty and risk abound, making Murphy's law a favorite engineering motto: if anything can go wrong, it will. Reminders of what can go wrong litter the history of engineering, and arguably they are as important as successes in guiding technological development (Petroski 1992). Horrific failures frequently become iconic, as with the *Titanic* sinking on its maiden voyage, killing over fifteen hundred passengers, in part due to a failure to have enough life rafts on what was believed to be an unsinkable ship. Many more people died after the nuclear reactor explosion at Chernobyl, especially from radiation-induced cancers. Such disasters are routinely called "accidents," but frequently they are foreseeable, at least as possibilities, and the damage caused could have been mitigated by greater mindfulness. Engineers need to be mindful of possible danger in everything from their original design ideas to the quality of the materials they use, and from the stresses placed on completed projects over time to possible misuse by consumers. Additional uncertainties surround the impact of engineering projects on the environment. Engineers' expertise includes the ability to proceed with incomplete knowledge—to proceed cautiously and conscientiously, taking into account myriad important factors that interact in complicated ways. It also demands reasoning combined with imagination in foreseeing and trying to counter what might go wrong.

The social experimentation model of engineering highlights mindfulness about informed consent. In medical ethics, patients' rights to give informed consent (and refusal) are front and center—at least they have become so in light of the patients' rights movement, combined with countless lawsuits arising from the failure of healthcare providers to convey to patients information about known risks. But with technological products and services, providers often avoid securing informed consent except where it is required by law. For example, social media services are accompanied by hopelessly elaborate or obscure cautions buried on websites that encourage users to sign up by clicking a computer key. In this and countless other ways, "our apps use us as much as we use our apps" (Turkle 2021, 341).

In addition to highlighting risk, the social experimentation model conveys the excitement, adventure, and creative possibilities in engineering. It highlights knowledge of new ways to advance human well-being. It calls for the openness to new information and perspectives that Langer's flexibility-mindfulness emphasizes, while grounding that openness in moral responsibilities that she too little appreciates. It celebrates the personal enjoyment and meaning highlighted in Csikszentmihalyi's flow-mindfulness, and what Samuel Florman (1994) famously called the existential pleasures of engineering. Above all, the model grounds virtue-mindfulness in conscientious and excellent engineering in the exercise of due care, diligence, and attention to detail. It embraces robust concepts of virtue-mindfulness, such as Buddhist—and Christian, Jewish, Muslim, and humanist—ideals of peace, compassion, and humanitarian love.

Engineers are often singularly well trained and well placed to foresee the dangers and creative possibilities of technology. Yet they are by no means solely responsible for the social experiments they participate in. Responsibility is shared with managers, technical staff, government inspectors, and users of technology. It is shared by the public, not only as consumers in purchasing and using technology but also as citizens.

Citizens

More than a legal status, citizenship implies participation in one's country. In addition, cosmopolitan notions of citizenship reach beyond nationalism to include global citizenship. Virtue-mindfulness concerning technology includes moral attunement to what matters economically; for example, poverty, the growing wealth gap, and socially responsible investment in the stock market. It also includes responsible attention to environmental concerns such as climate change, pollution, and the sustainable use of natural resources. And it includes political concerns about cyber terrorism and nuclear threats. I conclude with a few remarks on the macro dimensions of mindfulness about living in a technological world.

Psychologists tend to reinforce the view that moral values are merely subjective preferences, and their concepts of mindfulness have two additional limitations in understanding mindful technology at the macro level of citizenship.

First, their concepts tend to narrow how values are applied by emphasizing the immediate present—attention *to* the present, as well as attention *in* the present. Thus, Kabat-Zinn's value-suspending mindfulness-based meditation specifies focusing on what we are experiencing in the moment, Langer's flexibility-mindfulness centers on solving problems in our immediate situation, and Csikszentmihalyi's flow-mindfulness focuses on present activities. Of course, the present is a vague and elastic notion. It might refer to the moments in which we are meditating and focusing on our breathing, or instead to sustained activities involving flow. But it can also refer to periods of life, even long periods such as college years, young adulthood, and retirement—and sometimes it matters that we focus on such larger periods of time. In contrast with the short time horizons employed in psychological concepts of mindfulness, virtue-mindfulness requires integrating short-term focus with wider horizons of time and value, and skills in shifting between the two in appropriate ways. Macro themes about mindful technology remind us of the dangers of tunnel, silo, and blinkered vision by remembering the point of virtue-mindfulness: to promote good lives, good practices, and good communities. Useful conceptions of mindful technology will remind us of the bigger picture of living in a technological world and allow us to concentrate on what matters in our immediate situation.

Second, therapeutic practices of mindfulness, especially in connection with meditation, emphasize turning inward toward the self more than outward toward the world. In a therapeutic culture such as ours, this inward emphasis can easily reinforce self-absorption and narcissism. A familiar defense against this criticism is to argue that this excessive and narrow inward emphasis is a distortion of "true" or authentic mindfulness. Thus, Thomas Joiner, himself a psychologist, devotes an entire book to reclaiming genuine mindfulness as a proper blend of inward and outward attention: "Authentic mindfulness (moment-to-moment non-judgmental awareness of one's environment and subjective state) has lost its way, derailed by a culture of self-importance. It is turning into a vehicle for solipsism" (Joiner 2017, 39). He "pleads for a return to a selfless and authentic mindfulness, combined with a selfless stoicism gazing outwardly" (Joiner 2017, 185). I admire Joiner's book, but it would be more compelling if he had developed a concept of virtue-mindfulness, rather than borrowing Kabat-Zinn's value-suspending mindfulness.

Buddhist critics are on firmer ground in critiquing the influence of psychological concepts of mindfulness. They argue that psychologists degrade right mindfulness: "Rather than applying mindfulness as a means to awaken individuals and organizations from the unwholesome roots of greed, ill will and delusion, it is usually being refashioned into a banal, therapeutic, self-help technique that can actually reinforce those roots" (Purser and Loy 2013). Many people find Buddhist right mindfulness illuminating in thinking about technology and economics (Schumacher 1973; Badiner 2002). Certainly Buddhism's theme of conscientiously paying attention to and heeding moral responsibilities, and other moral considerations, is applicable to participating in a technological world (Vallor 2016). Yet Buddhism takes many forms, and so do Buddhist values and views about how best to apply them. For example, traditional Buddhism emphasizes minimizing the desires that attach us to the world in ways that cause suffering; hence it tends to be critical of capitalism, with its relentless creation of new desires for material goods. But other versions more readily accommodate the requirements of modern life. In any case, Buddhist concepts of right mindfulness are most promising for a pluralist society when they are stripped of much Buddhist orthodoxy (Harrison 2017).

Whether tied to a religious tradition or to a humanistic and secular value perspective, mindfulness about technology at the level of citizenship is always rooted in values and in perspectives on good lives. We are mindful as citizens insofar as we are attuned to what matters most about technology that affects others rather than merely ourselves. Such mindfulness might be shown in biking, walking, or using public transportation when possible, rather than driving a gas-guzzling car. It includes voting for politicians in light of their stands on funding science and education, rather than merely on personal pocketbook issues. It includes paying attention to issues about personal privacy in social media. Humility is needed, of course. All of us are enormously limited in how much we can know, or even be interested in. We quickly become overwhelmed and must rely on the advice of experts. Cautious optimism is also invaluable, despite the many dangers we face. Citizen-mindfulness at least extends to awareness of major issues that should concern us all. In this connection we might speak of shared or collective mindfulness.

In this chapter I have touched on only a few topics concerning mindfulness and technology. In doing so, I relied mainly on print sources at a time when the internet buzzes with blogs and additional discussions of mindfulness. Certainly much more needs to be said. I have tried to make the case for keeping central the idea of mindfulness as paying attention to what matters in light of relevant values. Mindfulness is a virtue insofar as one's values are sound. Mindful technology, then, is the use and creation of technology guided by paying attention to what matters in light of sound values. Hence the crucial topic is always which values are relevant and sound, a topic that psychological concepts of mindfulness often camouflage. The values include personal preferences but are not reducible to them, and any detailed specification of the relevant values will be subject to controversy but also dialogue and argument. The concept of virtue-mindfulness is not a magical solution to technological challenges, but it reminds us of the need for wisdom and moral responsibility in developing and using technology.

References

André, Christophe. 2011. *Looking at Mindfulness*. New York: Blue Rider Press.
Aristotle. 1976. *Nicomachean Ethics*. Rev. ed. Translated by J. A. K. Thomson and H. Tredennick. Harmondsworth, UK: Penguin.
Badiner, Allan Hunt. 2002. *Mindfulness in the Marketplace*. Berkeley, CA: Parallax Press.
Bodhi, Bhikku. 2013. "What Does Mindfulness Really Mean? A Canonical Perspective." In *Mindfulness: Diverse Perspectives on Its Meaning, Origins and Applications*, edited by J. Mark G. Williams and Jon Kabat-Zinn, 19–38. London: Routledge.
Boden, Margaret A. 2004. *The Creative Mind*. 2nd ed. New York: Routledge.
Collins, Mary Ann, and Teresa M. Amabile. 1999. "Motivation and Creativity." In *Handbook of Creativity*, edited by Robert J. Sternberg, 297–312. New York: Cambridge University Press.Csikszentmihalyi, Mihaly. 1990. *Flow: The Psychology of Optimal Experience*. New York: HarperPerennial.
Csikszentmihalyi, Mihaly. 1996. *Creativity: Flow and the Psychology of Discovery and Invention*. New York: HarperCollins Publishers.
Dewey, John. 1960. *Theory of the Moral Life*. New York: Holt, Rinehart, and Winston.
Doris, John M. 2002. *Lack of Character: Personality and Moral Behavior*. New York: Cambridge University Press.
Epstein, Ronald. 2017. *Attending: Medicine, Mindfulness, and Humanity*. New York: Scribner.

Florman, Samuel C. 1994. *The Existential Pleasures of Engineering*. 2nd ed. New York: St. Martin's Griffin.

Gelles, David. 2016. *Mindful Work*. New York: Mariner Books.

Gethin, Rupert. 2013. "On Some Definitions of Mindfulness." In *Mindfulness: Diverse Perspectives on Its Meaning, Origins and Applications*, edited by J. Mark G. Williams and John Kabat-Zinn, 263–79. London: Routledge.

Goleman, Daniel. 2013. *Focus: The Hidden Driver of Excellence*. New York: HarperCollins.

Gunatillake, Rohan. 2017. *Modern Mindfulness*. New York: St. Martin's Griffin.

Hadot, Pierre. 1995. *Philosophy as a Way of Life: Spiritual Exercises from Socrates to Foucault*. Translated by Michael Chase. Edited by Arnold I. Davidson. Oxford: Blackwell.

Harari, Yuval Noah. 2018. *21 Lessons for the 21st Century*. New York: Spiegel and Grau.

Harrison, Eric. 2017. *The Foundations of Mindfulness*. New York: The Experiment.

Joiner, Thomas. 2017. *Mindlessness: The Corruption of Mindfulness in a Culture of Narcissism*. New York: Oxford University Press.

Kabat-Zinn, Jon. 2013. *Full Catastrophe Living*. Rev. ed. New York: Bantam Books.

Kabat-Zinn, Jon. 2014. *Wherever You Go, There You Are: Mindfulness Meditation in Everyday Life*. 10th anniversary edition. New York: Hachette Books.

Kieran, Matthew. 2014. "Creativity as a Virtue of Character." In *The Philosophy of Creativity*, edited by Elliot Samuel Paul and Scott Barry Kaufman, 125–44. New York: Oxford University Press.

Langer, Ellen J. 2014. *Mindfulness*. 25th anniversary edition. Boston: Da Capo Press.

Langer, Ellen J. 2006. *On Becoming an Artist: Reinventing Yourself Through Mindful Creativity*. New York: Ballantine Books.

Levy, David M. 2016. *Mindful Tech: How to Bring Balance to Our Digital Lives*. New Haven: Yale University Press.

MacIntyre, Alasdair. 2007. *After Virtue*. Notre Dame, IN: University of Notre Dame Press.Martin, Mike W. 2000. *Meaningful Work: Rethinking Professional Ethics*. New York: Oxford University Press.

Martin, Mike W. 2006. *From Morality to Mental Health: Virtue and Vice in a Therapeutic Culture*. New York: Oxford University Press.

Martin, Mike W. 2007. *Creativity: Ethics and Excellence in Science*. Lanham, MD: Lexington Books.

Martin, Mike W. 2020. *Mindfulness in Good Lives*. Lanham, MD: Lexington Books.

Martin, Mike W., and Roland Schinzinger. 2005. *Ethics in Engineering*. 4th ed. Boston: McGraw-Hill.

Nozick, Robert. 1989. *The Examined Life*. New York: Simon and Schuster.

Penman, Danny. 2015. *Mindfulness for Creativity*. London: Little, Brown.

Peterson, Christopher, and Martin E. P. Seligman, eds. 2004. *Character Strengths and Virtues: A Handbook and Classification*. New York: Oxford University Press.

Petroski, Henry. 1992. *To Engineer Is Human: The Role of Failure in Successful Design*. New York: Vintage Books.

Purser, Ron, and David Loy. 2013. "Beyond McMindfulness." *Huffington Post*, July 1.

Rieff, Philip. 1987. *The Triumph of the Therapeutic*. Chicago: University of Chicago Press.

Schumacher, E. F. 1973. *Small Is Beautiful*. London: Blond and Briggs.

Siegel, Ronald D. 2010. *The Mindfulness Solution: Everyday Practices for Everyday Problems*. New York: Guilford Press.

Slote, Michael. 1983. *Goods and Virtues*. Oxford: Clarendon Press.

Thoreau, Henry D. 2014. *Walden*. Princeton: Princeton University Press.

Tiles, Mary, and Hans Oberdiek. 1995. *Living in a Technological Culture*. New York: Routledge.

Turkle, Sherry. 2021. *The Empathy Diaries*. New York: Penguin Press.

United Nations. 1946. *Constitution of the World Health Organization*. Geneva: World Health Organization.

Vallor, Shannon. 2016. *Technology and the Virtues: A Philosophical Guide to a Future Worth Wanting*. New York: Oxford University Press.

Weisberg, Robert W. 2006. *Creativity: Understanding Innovation in Problem Solving, Science, Invention, and the Arts*. Hoboken, NJ: John Wiley and Sons.

Zagzebski, Linda Trinkaus. 1996. *Virtues of the Mind: An Inquiry into the Nature of Virtue and Ethical Foundations of Knowledge*. New York: Cambridge University Press.

6

Virtuous Engineers

Ethical Dimensions of Technical Decisions

Jon Alan Schmidt

Introduction

The ancient Greeks generally identified three different kinds of knowledge: *episteme*, knowledge-that something is the case; *techne*, knowledge-how to achieve a predetermined outcome; and *phronesis*, knowledge-how to behave in a manner that is contextually sensitive and appropriate. Each of these respectively pertains to a specific sphere of human activity: *theoria*, which is contemplation or thinking; *poiesis*, which is production or making; and *praxis*, which is (inter)action or doing.

As such, *episteme*, *techne*, and *phronesis* also correspond with certain forms of rationality and judgment. However, for the sake of clarity, they will be translated here as theoretical knowledge, technical rationality, and practical judgment, respectively. These terms succinctly capture the fundamental aspects of each category, and defining them in parallel helps to highlight the key distinctions among them.

Theoretical knowledge is propositional in nature and aims at eternal truth. It consists of conceptual beliefs that count as facts when possessed by a person of understanding, who is characterized as intelligent and makes decisions based on evidence grounded in data. It applies primarily in the mental realm, resides in one's memory, and is imparted to a student by a process of instructing.

Technical rationality is procedural in nature and aims at external success. It consists of instrumental abilities that count as proficiencies when possessed by a person of skill, who is characterized as competent and makes decisions based on method grounded in rules. It applies primarily in the physical realm, resides in one's habits, and is imparted to an apprentice by a process of training.

Practical judgment is personal in nature and aims at internal integrity. It consists of ethical dispositions that count as virtues when possessed by a person of wisdom, who is characterized as prudent and makes decisions based on intuition grounded in experiences. It applies primarily in the social realm, resides in one's conscience, and is imparted to a disciple by a process of mentoring.

Jon Alan Schmidt, *Virtuous Engineers* In: *Science, Technology, and Virtues*. Edited by: Emanuele Ratti and Thomas A. Stapleford, Oxford University Press. © Oxford University Press 2021. DOI: 10.1093/oso/9780190081713.003.0007

Scholars have observed that today's culture has largely collapsed theoretical knowledge and practical judgment into technical rationality, with the result that technical rationality is widely regarded as the *only* legitimate form of reasoning. Its allure comes from its perceived objectivity and the apparent mastery over matter that humans have accomplished by employing it, as exemplified by technology. Because of this, "it is no longer seen as *a* form of rationality, with its own limited sphere of validity, but as coincident with *rationality as such*" (Dunne 2005, 374).

The two major types of modern ethical theories are grounded in technical rationality and are largely concerned with a person's outward behavior. Deontology prescribes adherence to particular rules or fulfillment of particular duties or obligations. An engineering code of ethics represents a clear application. Consequentialism evaluates a morally significant action on the basis of its actual or expected outcomes. Society frequently judges engineers in this manner, subjecting them to criticism and liability whenever failures occur.

On the surface, "technical rationality" sounds like the kind of thinking that engineers presumably use all the time. The popular perception, even among engineers themselves, is that engineering is precisely the rational solution of technical problems. There is indeed a rightful place for technical rationality, but it is only truly adequate to the task when the assignment at hand consists of following a detailed series of steps in order to achieve an already specified outcome.

Something more is necessary when life inevitably "present[s] us with a problematic situation where there is no discrete problem already clearly labelled as such, so that we might better speak of a difficulty or predicament rather than a problem." When confronting such circumstances, "one is not calculating the efficiency of different possible means towards an already determined end. Rather, one is often deliberating about the end itself: about what would count as a satisfactory, or at least not entirely unacceptable, outcome to a particular 'case' " (Dunne 2005, 381).

This observation is relevant to both engineering and ethics, because the formulation of such problems and their solutions is inherently indeterminate, routinely involving the selection of a way forward from among multiple options when no one "right" answer exists (Addis 1997). Intentionality, rather than rationality, is the operative conscious process (Schmidt 2013). Furthermore, even from a technical standpoint, engineering involves reconceptualizing a complex situation to facilitate analysis. In other words, it includes problem *definition*, not just problem *solution*.

Engineering Practice

In this respect, the "logic of ingenuity" routinely employed by engineers is analogous to the "logic of inquiry" routinely employed by scientists. The latter is typically prompted by *doubt*, "an uneasy and dissatisfied state from which we struggle to free ourselves and pass into the state of belief," which by contrast "is a calm and satisfactory state which we do not wish to avoid" (Peirce 1992, 114). Similarly, innovation is typically prompted by *dissatisfaction*: "the form of made things is always subject to change in response to their real or perceived shortcomings, their failures to function properly" (Petroski 1992, 22).

The scientific method can be summarized as involving three stages: *retroduction*, which is formulating an explanatory hypothesis, often in response to surprising observations; *deduction*, which is explicating the necessary consequences of the hypothesis in order to make predictions; and *induction*, which is conducting experiments in order to ascertain whether the hypothesis is falsified (Peirce 1998, 440–42). The retroductive aspect of engineering is conceiving a potential artifact that seems likely to fulfill a specified purpose, its deductive aspect is describing the artifact so that it can be physically made, and its inductive aspect is testing the artifact to confirm that it performs as expected (Schmidt 2016a).

This process is routine for *products* that will be mass-produced, but what about non-prototypical *projects*, each of which is one of a kind? In such cases, engineers must perform another cycle of retroduction-deduction-induction, commonly called analysis. The retroductive step is developing an idealized model of the proposed artifact and its immediate environment, the deductive step is processing this diagrammatic representation in accordance with idealized assumptions, and the inductive step is interpreting the results by comparing them with idealized rules. In other words, engineers must solve *real* problems by analyzing *fictitious* ones (Schmidt 2016b).

How do they accomplish this? One answer is that they use heuristics, defined as "anything that provides a plausible aid or direction in the solution of a problem but is in the final analysis unjustified, incapable of justification, and potentially fallible" (Koen 2003, 28). Classic examples of engineering heuristics include rules of thumb and factors of safety. Why do structural engineers consider the stress in a material that corresponds to 0.2 percent strain to be its yield strength and base most of their designs on this value? Why do they then typically divide it by 1.67 to determine a member's capacity to resist service loads? The short answer to both questions is: because it usually works.

In other words, heuristics cannot be "proven" in the absolute sense, but their utilization is warranted, frequently on the grounds of successful past implementation. In fact, each engineer has a unique collection of relevant heuristics at his or her disposal. When these heuristics are combined to facilitate "causing the best change in a poorly understood situation within the available resources" (Koen 2003, 7), they constitute a "design procedure," although this term is somewhat misleading because the outcome is not inevitable. On the contrary, "it is possible to produce very similar structural designs using different design procedures," and "similar design procedures can lead to significantly different structures—there is no logical connection between the two" (Addis 1990, 45–46).

What, then, guides an engineer to select the appropriate heuristics in a given set of circumstances? Over time, engineers develop conscious and unconscious ways of decomposing and solving problems based on what has and has not worked for them before. They may not be able to communicate these methods in words, because the methods have become integral to who they are and how they operate. Competence is achieved when an engineer can focus intuitively on what really matters—in particular, discerning the significant relations among the parts of an artifact, which must be captured in the analytical model of it (Schmidt 2016c)—and then converge relatively quickly on a viable solution.

Engineering Ethics

As an ancient alternative to modern theories, virtue ethics focuses on the person who acts, rather than the action itself. The emphasis is on *being* good, not just *doing* good: cultivating appropriate habits of conduct, rather than seeking to identify specific actions that are right in individual cases. With that in mind, there are at least three reasons virtue ethics is preferable to deontology and consequentialism for grounding and guiding the application of ethics to engineering.

First, the latter theories both are algorithmic in nature and attempt to impose universal principles that are supposed to govern actions in every situation, while virtue ethics is more heuristic and focuses on developing attitudes. Again, engineering is itself heuristic and non-deterministic, requiring creativity and skill to make choices from among multiple options. If engineering practice cannot be reduced to merely following a set of rules, then surely the same is true of engineering ethics.

Second, deontological and consequentialist ethics are "preventive," with a negative orientation; in an engineering context, they are largely geared toward avoiding professional misconduct and technological disasters, respectively. By contrast, virtue ethics is "aspirational," with a positive orientation; it advocates "the use of professional knowledge to promote the human good" (Harris 2008, 154).

Third, engineers are in constant danger of forgetting that their primary goal is not technical ingenuity for its own sake but helping people: "At its best, engineering changes the world for the benefit of humanity" (Bowen 2010, 135). However, both space and time often separate engineers from those affected by their work, who are thus perceived as populations rather than as individuals. Moreover, accountability is commonly diffused within an organization, rather than vested in a single person. Impersonal ethical systems such as deontology and consequentialism only exacerbate these disconnects.

Most contemporary proponents hold that virtues can only be properly identified within the context of a particular "practice" (MacIntyre 1981), which must have a significant social dimension as well as internal goods that are pursued as ends in themselves (see Stapleford and Hicks, Chapter 2 in this volume, for further discussion on MacIntyre's concept of a practice). However, the reputation of the engineering profession is such that most people likely assume that it does not meet either of these requirements. The common perception, even among engineers, is that engineering is primarily a matter of technical problem-solving and design by solitary individuals, and that the chief function of an engineer is to devise the most efficient means to achieve an end specified by someone else.

Trevelyan (2010, 175) challenges the first of these misconceptions, instead characterizing engineering as fundamentally a "human social performance" that "relies on harnessing the knowledge, expertise and skills carried by many people, much of it implicit and unwritten knowledge. Therefore social interactions lie at the core of engineering practice." Moreover, it is "a combined performance" carried out by a wide variety of stakeholders in which "the engineer's role is both to compose the music and conduct the orchestra" (Trevelyan 2010, 188). Music is a helpful metaphor, because composers and conductors provide definitive guidance to those who actually play the instruments but cannot dictate how they do so once the baton is raised.

As for the second misconception, *internal* goods are precisely those ends that are specific to a practice, can only be fully understood by those who participate in that practice, and generally benefit the entire practicing community. A virtue is then "an acquired human quality the possession and exercise of which tends to enable us to achieve those goods which are internal to practices and the lack of which effectively prevents us from achieving such goods" (MacIntyre 1981, 191). By contrast, *external* goods can be attained in a variety of ways, including different practices, often involving competition that leaves both winners and losers. Familiar examples include money, power, and status.

Although any practice requires a set of technical skills and the existence of institutions to sustain it over time, it is identical to neither of these. Every practice has its own history that goes beyond merely improving technical skills and serves as a tradition from which anyone who enters it must learn. Institutions are generally concerned with acquiring and distributing external goods, which is why virtues are so important: Without them, the institutions' pursuit of external goods would eventually supplant the practice's pursuit of internal goods, corrupting and ultimately destroying the practice (MacIntyre 1981).

Among today's professionals, this relationship between practices and institutions is especially relevant to engineers. Technology and innovation are generally dominated by market-driven value judgments, rather than technical knowledge. Even when managers or clients are engineers by training, the decisions that they make inevitably reflect the agendas and priorities of the organizations that they serve, but not necessarily the capabilities and limitations of the engineers whom they supervise or retain. As a result, engineering tends to be instrumental in nature: it is exploited by non-engineers as a convenient means of achieving their own objectives, which may be quite arbitrary (Goldman 1991).

Engineers might be able to escape this "social captivity," at least partially, by focusing on the specific kinds of goods that are internal to their practices and "whose intended achievement defines them as the particular practices that they are." Such internal goods include "both competencies proper to each practice and virtues of character that transcend any particular practice" (Dunne 2005, 368). The latter are essential to ensure that the former are not treated merely as means for acquiring external goods, such that "the practice can be made instrumental to the point that violation of its internal fabric is allowed" (Dunne 2005, 369). A practice is thus "something that can succeed or fail in being true to its own proper purpose" (Dunne 2005, 367).

Proper Purpose

What does it mean for a practice to have a proper purpose? Miller (1984, 51) suggests that there are two kinds of practices:

> There is an important distinction to be drawn between practices which have no *raison d'etre* other than the particular excellences and enjoyments which they allow to participants (I shall refer to such practices as "self-contained") and practices which have a wider social purpose (I shall refer to these as "purposive"). Games, from which much of MacIntyre's thinking about practices seems to be drawn, are the main exemplars of the first category.... On the other hand, in the case of a productive activity ... there is an external purpose which gives the practice its point and in terms of which it may be judged.

In other words, MacIntyre's rigid classification of all goods as either internal or external to any given practice, as well as the presumptive exclusion of "productive activity" (*poiesis*) from the scope of a genuine practice (*praxis*), is difficult to maintain if the practice is one that could be described as purposive (Stapleford 2018). Such a practice has an end that is intrinsic in a way that a strictly external good cannot be; and yet that end also does not qualify as an internal good, since its benefits extend well beyond the boundaries of the practicing community.

Mitcham (2009) argues that engineering has no such ideal that is "good in itself" and is well embedded in its curriculum and practice, characterizing the profession as "philosophically inadequate" on that basis. Although engineers are explicitly charged with holding paramount the safety, health, and welfare of the public, he points out that they are not especially qualified to determine exactly what satisfies that obligation. Goldberg (2009) responds that Mitcham is merely reacting to the ethical complexity of engineering. He acknowledges that strictly serving a manager's or client's interests will not always align naturally with the interests of society as a whole, but it does not follow that engineering does not have a proper purpose *at all*.

The ancient Greeks insisted that the proper purpose of *any* worthwhile activity is to facilitate some aspect of *eudaimonia*, a word that is usually equated with "happiness" but is best translated as "well-being" or "human flourishing." Favorable physical, social, and material conditions can be important factors for living a genuinely good life, and various practices are engaged in providing and maintaining such conditions. With this in mind, Bowen (2010, 142) describes the end of engineering as "the promotion of human flourishing through contribution to material well-being."

Moreover, it is important to stipulate that purposive practices should foster *universal* well-being; hence engineers ought to work toward the material well-being of *all* people, not just a privileged group (Miller 1984, 55):

> If a "practice" account of the virtues is going to be successful, the practices concerned must be those I have called purposive, and moreover those whose aims are fairly central to human existence. By implication it is a mistake to try to explain the virtues by reference to goods *internal* to practices. Although MacIntyre is quite right to draw our attention to the existence of such goods— for even in the case of purposive practices standards of excellence will develop whose achievement will be regarded as an internal good by the participants— the virtues themselves must be understood in relation to those wider social purposes which practices serve.

The proper purpose of a purposive practice thus constrains what will count as its internal goods, since those goods must tend to advance that purpose in some way. Likewise, the personal traits that enable someone to achieve the internal goods of a purposive practice only qualify as virtues if they also facilitate the accomplishment of its proper purpose. Consequently, the internal goods and virtues that are specific to the practice of engineering must be ends and attributes, respectively, that uniquely contribute to the material well-being of all people.

Societal Role

What exactly is "material well-being" in this context? How do engineers bring it about? For one thing, "Engineering projects provide us with the technological means of overcoming some of the physical limitations that are a consequence of being human." Engineering is thus "a profession that seeks to harness techno-logical advancements to provide solutions to a wide range of social problems." Because of this, there is a specific aspect of engineering that is crucial to under-standing the peculiar ethical burden that engineers bear (Ross and Athanassoulis 2010, 148):

> Engineering projects are often innovative, long-term and involve the co-ordination of so many different variables that it is impossible to predict abso-lutely accurately what their consequences will be. In addition, because of the scale, and infra-structural nature of these projects there is often significant po-tential to do harm should something go wrong.

The engineer thus assumes a responsibility to determine which dangers are per-tinent to each undertaking, decide how best to deal with them in spite of the "three enemies of knowledge"—ignorance, uncertainty, and complexity—that are associated with them (Elms 1999, 321–22), and inform everyone who needs to become aware of them. In other words, the basic societal role of engineering is the assessment, management, and communication of risk: the very real pos-sibility that engineered products and projects could *detract* from the material well-being of *some* people, rather than *enhancing* the material well-being of *all*.

The concepts of risk and responsibility, as well as the relationship between them, can be somewhat ambiguous. For example, Hansson (2009, 1069–71) outlines at least five relevant senses of the word *risk*:

1. A harm that may or may not occur
2. The cause of a harm that may or may not occur
3. The probability of a harm that may or may not occur
4. The statistical expectation value of a harm that may or may not occur
5. The fact that a decision is made when the outcome probabilities are known

The first three definitions are often assigned to alternative terms—such as *consequence, hazard,* and *threat* or *vulnerability,* respectively—and treated as contributors to the fourth, which is the standard technical meaning of *risk* among engineers and risk analysts. The fifth is only invoked in formal decision theory, where "decision under risk" is an alternative to "decision under certainty," when each option is associated with exactly one outcome, and "decision under uncertainty," when outcome probabilities are unknown. All five acknowledge the indeterminate nature of risk: even when quantified, the concept reflects a degree of belief, rather than a fact (Elms 1999, 313).

The idea of risk is even more nuanced in fields such as psychology and other social sciences. Contextual elements come into play, which may include "by whom the risk is run, whether the risk is imposed or voluntary, whether it is a natural or man-made risk, and so on" (Van de Poel and Nihlén Fahlquist 2012, 881). Such factors imply at least some degree of responsibility on the part of those who confront it on behalf of others, including engineers. Van de Poel (2011) identifies at least nine distinct notions of "responsibility" in common usage:

1. Cause: The earthquake was responsible for killing 100 people.
2. Role: The driver is responsible for controlling the vehicle.
3. Authority: The superintendent is responsible for the construction project.
4. Capacity: The person has the ability to act in a responsible way.
5. Virtue: The person has the disposition to act in a responsible way.
6. Obligation: The person is responsible for the safety of the passengers.
7. Accountability: The person is responsible for explaining what he/she did.
8. Blameworthiness: The person is responsible for what happened.
9. Liability: The person is responsible for the cost of the damages.

The first four are strictly descriptive, while the other five are normative; of those five, virtue and obligation are forward-looking (prospective), while the last three are backward-looking (retrospective). Philosophers throughout history have mostly addressed retrospective responsibility, especially blameworthiness—in particular, the necessary and sufficient conditions for someone to be properly held responsible for something that has already occurred. However, engineering ethics is—or at least should be—more concerned with the prospective responsibility of engineers who assess, manage, and communicate risk in an effort to avoid harms.

Internal Goods

Admittedly, "it would be convenient if there was a formula for making good and right decisions about whether, when, and what to risk" (Ross and Athanassoulis 2012, 834)—which is essentially what the modern ethical theories purport to offer: definitive guidance derived from universal principles, such as assumed duties and obligations in the case of deontology, or assessment of potential outcomes in the case of consequentialism. By contrast, virtue ethics recognizes that any truly substantive ethical inquiry will lead to "a complex, varied, and imprecise answer that cannot be captured in an overriding rule" (Ross and Athanassoulis 2012, 836).

For one thing, people participate in any instance of risk-taking in one or more of three ways: as the decision-maker, as the potential harm-bearer, or as the intended beneficiary (Athanassoulis and Ross 2010, 150–51). It is not morally problematic when a single person occupies all three positions, but for engineering risks, multiple parties are always implicated: the engineer makes the decision, the public is often in harm's way, and the engineer's employer or client presumably stands to gain something. Under such circumstances, what factors influence whether the risks associated with a given engineering decision—presumably encompassing the potential outcomes, their corresponding likelihoods, and relevant contextual features (Elms 1999, 313–15)—are reasonable, and therefore justifiable?

The widespread assumption that answering this question is purely a matter of "objective" probabilistic calculation is mistaken. Instead, a number of "subjective" considerations inevitably come into play, including the desires and priorities of the engineer, different perspectives on how to describe various outcomes should they come about, and the range of available options (Ross and Athanassoulis 2010, 153–57). Virtue ethics shifts the focus from individual actions to patterns of behavior: "choices that people make, those choices that are reaffirmed over time, and those choices that express their deeply held values and beliefs" (Ross and Athanassoulis 2012, 839).

The primary concern is thus with someone's long-term attitude toward risk, especially with respect to the potential impacts on the well-being, material and otherwise, of others. The central concept here is *character*, defined as "the set of stable, permanent, and well-entrenched dispositions to act in particular ways" (Athanassoulis and Ross 2012, 840, citing Athanassoulis 2005, 27–34). These dispositions qualify as virtues when they enable and incline someone "to respond well to whatever situation is encountered" (Ross and Athanassoulis 2012, 840), which usually entails having "a clear and accurate view of the situation" and producing "a proportionate, rational response" (Athanassoulis and Ross 2010, 225).

What transpires may not be entirely within the control of the person responding, so what matters from an ethical standpoint is the quality of the decision at the time when it is made, rather than the effects that emerge from it afterward. Therefore, "the assignment of moral responsibility for risk-taking and for the results of risk-taking needs to be done on a case by case basis" (Ross and Athanassoulis 2010, 157). Nevertheless, engineers share a common consensus, although they rarely articulate it, about what they do and how it fits into the bigger picture (Ross and Athanassoulis 2010, 159–60):

It is our contention that the chief good internal to the practice of engineering is safe efficient innovation in the service of human wellbeing and that this good can only be achieved where highly accurate, rational decisions are made about how to balance the values of safety, efficiency and ambition in particular cases. . . . [E]ngineers don't just strive to find technological solutions to human problems, they strive to do so in a manner fitting for the conduct of an engineer which involves consciously foregrounding the values of safety and sustainability.

This passage invokes the notion of an internal good and connects it directly with engineering's proper purpose. However, the references to "values" seem out of place, and the attempt to pinpoint a single internal good is needlessly restrictive. Instead, upon rearranging the terminology, three such goods emerge:

1. Safety: protecting people and preserving property
2. Sustainability: improving environments and conserving resources
3. Efficiency: performing functions while minimizing costs

These three distinct aspects of material well-being and risk mitigation are goals inherent to some degree in nearly every engineering endeavor today. Engineers can—and regularly do, even if only subconsciously—treat them as ends in themselves, rather than as means to some other end. They qualify as goods that are internal to the practice of engineering because they are specific to it, can only be fully understood as defined here by those who participate in it, and generally benefit the entire practicing community. Even so, it is important to acknowledge that safety, sustainability, and efficiency may be—and in fact, frequently are—in tension with each other to some extent.

Moral Virtues

The question then arises: What personal attributes would enable someone to make the necessary trade-offs among these internal goods without inappropriately compromising any of them? The first step is to admit that "the picture of engineering as morally neutral is misleading" (Busby and Coeckelbergh 2003, 364). It is true, as discussed earlier, that engineers are not completely autonomous and rarely have the authority to establish on their own the degree of risk that is acceptable for a given assignment. However, this does not absolve engineers from taking risk into account on a case-by-case basis, especially in its moral dimensions (Coeckelbergh 2006). Instead of only asking, "How can I justify the design that I want to develop?" engineers should also wonder, "How can I find the design that reasonably minimizes risk?" (Busby and Coeckelbergh 2003, 364).

There are at least three motives for engineers to embrace this kind of "ascribed ethics." First, people generally behave in accordance with their expectations, so understanding common presuppositions about engineers and (especially) engineered systems may help better inform the design process. Second, non-engineers perceive risks that engineers, precisely because of their specialized expertise, are less likely to notice. Third, and most important, "the ability to imagine the implications of one's actions, such as taking risks with others' welfare in one's product design . . . [is] as important to morality as any general principle" (Busby and Coeckelbergh 2003, 366).

In the *Nicomachean Ethics*, Aristotle (1984) advocates locating most virtues at the mean between corresponding extremes of excess and deficiency that are deemed to be vices. With this in mind, adopting the standpoint of those put at risk by engineering endeavors in order to identify the types of behavior that engineers ought to avoid may lead to insights about those to which they should aspire. It seems logical to begin with three specific virtues that are widely viewed as prerequisites for all practices: justice, courage, and honesty. Without them, internal goods of any kind are ultimately unattainable, because these characteristics are integral to the types of relationships that must be maintained among the participants in a practice (MacIntyre 1981, 191):

> It belongs to the concept of a practice as I have outlined it . . . that its goods can only be achieved by subordinating ourselves within the practice in our relationship to other practitioners. We have to learn to recognize what is due to whom; we have to be prepared to take whatever self-endangering risks are demanded along the way; and we have to listen carefully to what we are told about our own inadequacies and to reply with the same carefulness for the facts.

Justice is necessary because the authority of the standards of excellence that define a practice must be accepted by all who enter into it, along with the inadequacy of their initial performances when measured by those standards. Courage is necessary because the willingness to sacrifice is a component of genuine concern for others, and pursuing internal goods may sometimes require foregoing external goods. Honesty is necessary because trust is indispensable, not only among practitioners, but also between them and the general public, especially in a practice like engineering that involves significant uncertainties.

The picture becomes clearer when we recognize how these three virtues apply within the practice of engineering. Justice precludes both favoritism and indifference; every single person who will potentially be affected by what an engineer does deserves due consideration. Courage calls for being neither overconservative nor overconfident; engineers must "balance degrees of caution and (social) ambition that are appropriate to the circumstances and nature of [their] decisions" (Ross and Athanassoulis 2010, 157). Honesty means eschewing both deception and indiscretion; respect for confidentiality must be balanced with the public interest.

Moriarty (2008, 124–33) proposes a similar trio of virtues even more tailored to engineers. Objectivity is a stance of impartiality or fairness that diligently examines all relevant factors and ultimately resolves each matter on the merits. Care entails assuming personal concern for another and then instinctively doing whatever the situation demands. Honesty encompasses cooperation and transparency as well as truthfulness.

Of course, these attributes are hardly exclusive to engineering; perhaps there is nothing more to being a virtuous engineer than being a virtuous person who happens to be an engineer. While this is accurate to an extent, Miller's insight about purposive practices leads to the realization that Moriarty's three moral virtues of engineering align with the three components of its societal role: virtuous engineers objectively assess risk, carefully manage risk, and honestly communicate risk in order to achieve safety, sustainability, and efficiency for the sake of the material well-being of all people.

Intellectual Virtue

But how does this happen in spite of the potential for conflicts among the individual goods or virtues in each list? Many philosophers, going all the way back to the ancient Greeks, have drawn a strong analogy between virtues and skills (Stichter 2007), so the same terminology may be applied to engineers who characteristically exhibit objectivity, care, and honesty in the proper proportions. Ethical competence should be seen as an essential component of technical competence, such that it is by definition impossible for an engineer to be truly competent and unethical at the same time.

As engineers gain experience, they internalize practice-specific goods and virtues to the point that they are able to balance them rightly in particular cases, having developed "a reliable capacity to respond to risk with the appropriate attitude . . . [P]rofessionals acquire, through training and thought, settled dispositions to judge in accordance with their distinctive professional values and thus can be said to exemplify a kind of professional practical wisdom" (Ross and Athanassoulis 2010, 163)—in other words, phronesis. Such practical judgment appears to be closely related to a faculty that engineers constantly take for granted but rarely try to explain: engineering judgment. It is widely understood to be so critical that "one who otherwise knows what engineers know but lacks 'engineering judgment' may be . . . a handy resource much like a reference book or database, but cannot be a competent engineer" (Davis 2012, 789–90).

Judgment in this sense is "the *disposition* (including the ability) to *act* as competent members of the discipline act" (Davis 2012, 790). It involves more than just theoretical knowledge-that or even technical knowledge-how; it is "the embodiment of a high likelihood of making certain decisions in the appropriate way at the appropriate time" (Davis 2012, 798–99). Such judgment is neither arbitrary nor algorithmic, and the reference to peers as the benchmark is consistent with the legal notion of the standard of care: "that level or quality of service ordinarily provided by other normally competent practitioners of good standing in that field, contemporaneously providing similar services in the same locality and under the same circumstances" (Kardon 1999).

Judgment could also be seen as the key to integrating ethics into any discipline that requires it. "Once we see judgment as central to the discipline, we can also see how central ethics is to its competent practice. There is no good engineering, no good science, and so on without good judgment and no good judgment in these disciplines without ethics." However, there is still the matter of which approach to ethics is being employed. Davis reveals a deontological orientation: "I mean those (morally permissible) standards of conduct (rules, principles, or ideals) that apply to members of a group simply because they are members of that group. Engineers need to understand (and practice) engineering ethics to be good engineers, not moral theory, medical ethics, or the like" (Davis 2012, 790). This actually makes ethics a matter of technical rationality or *techne*, rather than practical judgment or phronesis.

Despite recognizing that engineering judgment, like phronesis, is "a disposition to act in an appropriate way," Davis (2012, 802–4) defines the latter much more broadly as "the ability reliably to respond to any situation with a course of action that makes life better . . . *Phronesis* is (more or less) a global term; judgment is not global . . . We should speak of the art, craft, or skill of [an engineer] rather than his *phronesis* when he shows good [engineering] judgment." Nevertheless, Davis (2012, 801) explicitly wonders whether engineering judgment is a virtue, since it admittedly "is a disposition that contributes to living well (both to the engineer's living well and to others living well)." Ultimately, though, he is worried about the limited scope of judgment in this sense: "The traditional virtues (courage, hospitality, truthfulness, and so on) concern the whole of life. No traditional virtue concerns only a single discipline as, for example, engineering judgment does."

Such a concern is misplaced once virtues are situated within distinct practices (MacIntyre 1981). Engineering judgment is, in fact, a discipline-specific form of practical judgment, which the ancient Greeks classified as an *intellectual* virtue: importantly, the one that guides and ultimately unifies the corresponding *moral* virtues.

Conclusion

The framework presented in this chapter for applying virtue ethics to engineering practice may be summarized under three headings that correspond to the central concepts in the classical approach.

Praxis: The "What" of Engineering

First, although engineering is a "productive activity," its social aspect is such that engineers engage in a combined human performance in which they play a particular societal role: the assessment, management, and communication of risk.

Engineers must convince others to hire or retain them, and then ascertain and attempt to satisfy those stakeholders' expectations for each assignment. Furthermore, research has repeatedly indicated that engineers across all disciplines, career stages, and types of employers spend the majority of their time at work interacting with others (Trevelyan 2010). Engineering is thus always a collaborative endeavor, assembling expertise that is distributed among multiple participants; someone whose technical activities are self-motivated and solitary is more accurately labeled as an inventor.

Engineers are the decision-makers in situations where members of the public are usually the potential harm-bearers, even when they are also supposed to benefit in some way. The latter take it for granted that engineering design adequately accounts for all of the applicable hazards, and thus ascribe to engineers the obligation to mitigate them. Embracing this responsibility entails not only recognizing these uncertainties and dealing with them appropriately but also calling attention—preferably beforehand—to any residual risk that is associated with an engineered product or project, including anticipated social and environmental impacts.

Phronesis: The "How" of Engineering

Second, engineers exercise the intellectual virtue of engineering, which is practical judgment—specifically, engineering judgment—while exhibiting the moral virtues of engineering, which are objectivity, care, and honesty.

Engineers routinely confront difficulties and predicaments, rather than well-structured problems that have deterministic solutions. Learning theories, rules, and maxims—that is, heuristics and design procedures—provides a necessary and solid foundation. However, it is only through experience that someone can develop the skill to discern quickly what is important in a specific set of circumstances and then select a suitable way forward.

Risk assessment requires objectively evaluating the likelihood and severity of possible threats, and then identifying alternatives for reducing one or both of these parameters. Risk management requires carefully deliberating over multiple viable options and then choosing one that rightly balances caution and ambition on behalf of all those who may be affected. Risk communication requires honestly acknowledging the dangers that cannot reasonably be eliminated, and then informing everyone who needs to be aware of them.

Eudaimonia: The "Why" of Engineering

Third, engineers strive to fulfill the proper purpose of engineering, which is to enhance the material well-being of all people, by achieving the internal goods of engineering, which are safety, sustainability, and efficiency.

It takes a deliberate decision and ongoing resolve to do this faithfully. Engineers must prioritize the proper purpose and internal goods of their practice not only over their own immediate interests but also over the external goods valued by those who make the major decisions and pay the bills. The prospective reward for doing so is the opportunity to escape, at least partially, the "social captivity" that renders engineering largely instrumental, subject to exploitation by managers and clients.

As a step in this direction, engineers can pursue their most fundamental aims—protecting people and preserving property, improving environments and conserving resources, and performing functions while minimizing costs—for their own sake, rather than merely as means to another end. When merged, these aims constitute an overall notion of quality that engineers seek to incorporate into everything that results from their efforts. Nelson (2012) hints at this by writing that "design is inherently goalless" because the precise outcome is unknown during the process of creating it.

Summary

Uniting all of these ideas in an arrangement that states what engineers do, how they do it, and why it matters in broad terms, and then presents the details in the reverse order, produces the following concise yet comprehensive statement of engineers' unique and vital contribution to human flourishing:

> Virtuous engineers assert their responsibility
> for engaging in a combined human performance
> that involves the exercise of practical judgment
> to enhance the material well-being of all people
> by achieving safety, sustainability, and efficiency
> while exhibiting objectivity, care, and honesty
> in assessing, managing, and communicating risk.

This formulation offers an aspirational vision of what it looks like for an engineer to practice with genuine integrity. Virtue ethics is less concerned with what someone has done and will do than with what kind of person—what kind of engineer—someone is now and will become in the future. The goal is not so much better engineering decisions, but better engineering decision-makers; that is, better engineers.

Moreover, adopting virtue ethics is unlikely to require engineers to change radically what they are already doing on a day-to-day basis. Deontology and consequentialism effectively treat ethics as something that engineers pursue separately from engineering itself. By contrast, a virtue perspective affirms that ethics is fully integral to the profession. In fact, ethical deliberation in any context often conforms to the "logic of ingenuity" that is familiar to engineers: faced with uncertainty about how to proceed, an individual must imagine possibilities, assess alternatives, and choose one of them to actualize (Schmidt 2017).

Acknowledgment

This chapter is largely an abridged version of the author's paper, "Changing the Paradigm for Engineering Ethics," *Science and Engineering Ethics*, 20(4), 985–1010, December 2014, DOI 10.1007/s11948-013-9491-y, and is reprinted by permission from Springer Nature. The final publication is available at http://link.springer.com/article/10.1007/s11948-013-9491-y.

References

Addis, W. 1990. *Structural Engineering: The Nature of Theory and Design*. New York: Ellis Horwood.

Addis, W. 1997. "Free Will and Determinism in the Conception of Structures." *Journal of the International Association for Shell and Spatial Structures* 38, no. 2: 83–89.

Aristotle. 1984. *The Complete Works of Aristotle: The Revised Oxford Translation.* Edited by J. Barnes. Princeton, NJ: Princeton University Press.

Athanassoulis, N. 2005. *Morality, Moral Luck and Responsibility.* Basingstoke: Palgrave Macmillan.

Athanassoulis, N., and A. Ross. 2010. "A Virtue Ethical Account of Making Decisions About Risk." *Journal of Risk Research* 13, no. 2: 217–30.

Bowen, R. 2010. "Prioritizing People: Outline of an Aspirational Engineering Ethic." In *Philosophy and Engineering: An Emerging Agenda,* edited by I. Van de Poel and D. Goldberg, 135–46. Dordrecht: Springer.

Busby, J., and M. Coeckelbergh. 2003. "The Social Ascription of Obligations to Engineers." *Science and Engineering Ethics* 9, no. 3: 363–76.

Coeckelbergh, M. 2006. "Regulation or Responsibility? Autonomy, Moral Imagination, and Engineering." *Science, Technology, and Human Values* 31, no. 3: 237–60.

Davis, M. 2012. "A Plea for Judgment." *Science and Engineering Ethics* 18, no. 4: 789–808.

Dunne, J. 2005. "An Intricate Fabric: Understanding the Rationality of Practice." *Pedagogy, Culture and Society* 13, no. 3: 367–89.

Elms, D. 1999. "Achieving Structural Safety: Theoretical Considerations." *Structural Safety* 21, no. 4: 311–33.

Goldberg, D. 2009. "Is Engineering Philosophically Weak? A Linguistic and Institutional Analysis." In *Proceedings, SPT 2009: Converging Technologies, Changing Societies,* 226–27. Twente: University of Twente.

Goldman, S. 1991. "The Social Captivity of Engineering." In *Critical Perspectives in Nonacademic Science and Engineering,* edited by P. Durbin, 125–52. Bethlehem, PA: Lehigh University Press.

Hansson, S. 2009. "Risk and Safety in Technology." In *Handbook of the Philosophy of Science: Philosophy of Technology and Engineering Sciences,* edited by A. Meijers, 1069–102. Oxford: Elsevier.

Harris, C. 2008. "The Good Engineer: Giving Virtue Its Due in Engineering Ethics." *Science and Engineering Ethics* 14, no. 2: 153–64.

Kardon, J. 1999. "The Structural Engineer's Standard of Care." Online Ethics Center for Engineering and Research, July 31. http://www.onlineethics.org/Topics/ProfPractice/PPCases/standard_of_care.aspx.

Koen, B. 2003. *Discussion of the Method: Conducting the Engineer's Approach to Problem Solving.* New York: Oxford University Press.

MacIntyre, A. 1981. *After Virtue.* Notre Dame, IN: University of Notre Dame Press.

Miller, D. 1984. "Virtues and Practices." *Analyse und Kritik* 6, no. 1: 49–60.

Mitcham, C. 2009. "A Philosophical Inadequacy of Engineering." *The Monist* 92, no. 3: 339–56.

Moriarty, G. 2008. *The Engineering Project: Its Nature, Ethics, and Promise.* University Park, PA: Pennsylvania State University Press.

Nelson, E. 2012. "A Structural Engineer's Manifesto for Growth, Part 1." *Structure* 19, no. 4: 74.

Peirce, C. 1992. *The Essential Peirce: Selected Philosophical Writings, Volume 1 (1867–1893).* Edited by N. Houser and C. Kloesel. Bloomington: Indiana University Press.

Peirce, C. 1998. *The Essential Peirce: Selected Philosophical Writings, Volume 2 (1893–1913).* Edited by the Peirce Edition Project. Bloomington: Indiana University Press.

Petroski, H. 1992. *The Evolution of Useful Things.* New York: Vintage Books.

Ross, A., and N. Athanassoulis. 2010. "The Social Nature of Engineering and Its Implications for Risk-Taking." *Science and Engineering Ethics* 16, no. 1: 147–68.

Ross, A., and Athanassoulis, N. 2012. "Risk and Virtue Ethics." In *Handbook of Risk Theory*, edited by S. Roeser, R. Hillerbrand, P. Sandin, and M. Peterson, 833–56. Dordrecht: Springer.

Schmidt, J. 2013. "Engineering as Willing." In *Philosophy and Engineering: Reflections on Practice, Principles, and Process*, edited by D. Michelfelder, N. McCarthy, and D. Goldberg. Dordrecht: Springer.

Schmidt, J. 2016a. "The Logic of Ingenuity, Part 1: Engineering Design." *Structure* 23, no. 9: 63.

Schmidt, J. 2016b. "The Logic of Ingenuity, Part 2: Engineering Analysis." *Structure* 23, no. 10: 46.

Schmidt, J. 2016c. "The Logic of Ingenuity, Part 3: Engineering Reasoning." *Structure* 23, no. 11: 62–63.

Schmidt, J. 2017. "The Logic of Ingenuity, Part 4: Beyond Engineering." *Structure* 24, no. 3: 66–67.

Stapleford, T. 2018. "Making and the Virtues: The Ethics of Scientific Research." *Philosophy, Theology and the Sciences* 5, no. 1: 28–50.

Stichter, M. 2007. "Ethical Expertise: The Skill Model of Virtue." *Ethical Theory and Moral Practice* 10, no. 2: 183–94.

Trevelyan, J. 2010. "Reconstructing Engineering from Practice." *Engineering Studies* 2, no. 3: 175–95.

Van de Poel, I. 2011. "The Relation Between Forward-Looking and Backward-Looking Responsibility." In *Moral Responsibility: Beyond Free Will and Determinism*, edited by I. Vincent, I. van de Poel, and J. van den Hoven, 37–52. Dordrecht: Springer.

Van de Poel, I., and J. Nihlén Fahlquist. 2012. "Risk and Responsibility." In *Handbook of Risk Theory*, edited by S. Roeser, R. Hillerbrand, P. Sandin, and M. Peterson, 877–907. Dordrecht: Springer.

7

Artificial Phronesis

What It Is and What It Is Not*

John P. Sullins

Phronesis and Robotics Technologies

Robotics technologies, especially social robotics technologies, are very likely to affect the development of human virtue and character in the coming decades. This chapter is written from the standpoint of virtue as it applies to information technologies (see Vallor, Chapter 4 in this volume, and Vallor 2016). While virtue ethics refers to the long, multicultural study of virtues and human character, we will not be able to linger there for long. In this chapter we are interested in a strange problem that never occurred to the early thinkers in virtue ethics—namely, how do technologies help build our human character, and how do we shape the character of our technologies to reflect our virtues rather than our vices? This is a big project, and here we will only take on a modest portion of the problem, specifically as it relates to the role that phronesis (practical wisdom) plays in the development of character in human agents and how we might mirror that capacity in robotic systems that are designed to interact with humans in ethically impactful ways.

Phronesis has many complex philosophical connotations, but to begin with, the term will be used here to refer to the "practical wisdom" needed by agents (human or otherwise) that is learned by acting in social situations and gives those agents that possess this quality the ability to make new or novel judgments. These judgments are sincere attempts at producing good and correct actions that lead to the best ethical outcomes possible in a given situation. Phronesis plays a role in a wide variety of human interactions, from the life-and-death decisions that a physician might make in a triage situation down to simple public interactions on the street.

Given that we have robots and AI assistants that are, or will soon be, involved in influencing life-and-death decisions in medicine and that social robots are machines that are designed to interact with humans in all manner of everyday social situations, we can see that roboticists envision a world where some robotics technology is interacting with us at all times. Many of those interfaces are ones

John P. Sullins, *Artificial Phronesis* In: *Science, Technology, and Virtues.* Edited by: Emanuele Ratti and Thomas A. Stapleford, Oxford University Press. © Oxford University Press 2021. DOI: 10.1093/oso/9780190081713.003.0008 that, if done by a human, would require a good deal of phronetic judgment. How then are we going to program the right phronetic judgment into our machines? Artificial phronesis (AP) is a term that was coined to refer to the various engineering solutions that attempt to solve the problem just described (Sullins 2019; Dietrich et al. 2021).

As a neologism, AP needs to be carefully conceived given that the standard definition of phronesis in terms of classical virtue ethics never imagined that it would be applied to anything other than human beings. We are self-consciously stretching the traditional conception of phronesis, but hopefully not to the breaking point. If it is true that phronesis is only possible in humans, then AP is a self-contradictory claim. That may be so, but here we will entertain the notion that it is at least an imaginable trait that machines might have. We imagine potential entities with this skill quite often in media. Ethically complex and socially aware behavior is found to various degrees in many of the popular robotic characters from fiction. It is, of course, acknowledged here that appeals to fiction are no proof of concept. For instance, many animals or mythological creatures display ethically complex and socially aware phronetic abilities on a regular basis in fiction, even though talking animals or mythological creatures are not commonly encountered in our reality. We can imagine a lot of things that will never be realized, but the appeal to fiction does show that we, as a species, do seem to desire that phronesis be more widely spread throughout our environment. That desire will naturally play out as we imagine a new future through our built technologies. We are engaged in a kind of applied science fiction when the design of high technology is done. Famously, the very word *robot* was invented in a speculative fiction play, and ever since robotics technology and science fiction have had close ties. Roboticists commonly cite science fiction as the inspiration for their work. As an example, the very first paragraphs in the preface to the groundbreaking book *Designing Sociable Robots* by the roboticist Cynthia Breazeal describe the deep inspiration she received as a child from the robot characters in the movie *Star Wars* and how that eventually led her to

> define a vision for sociable robots of the future. Taking R2-D2 and C-3PO as representative instances, a sociable robot is able to communicate and interact with us, understand and even relate to us, in a personal way. It is a robot that is socially intelligent in a human-like way. We interact with it as if it were a person, and ultimately a friend. This is the dream of a sociable robot. (Breazeal 2002, xi)

Those words were written seventeen years prior to this chapter, and as this chapter is being written Breazeal is having trouble bringing to market a sociable robot platform for the home based on her dreams. Her project has received many endorsements and initial investors, both private and corporate, who gave \$3.5 million to see this machine come into existence (Mitchell 2018). The resulting robot, named Jibo, while a fascinating piece of technology, fell prey to the realities of the market, and the project is on hold as of the time of this publication. Part of the problem of robots like Jibo is that our imaginations far outstrip the capabilities of these machines. Jibo looks like an amiable droid from science fiction, but it just does not have the same personality that is modeled by the robots of our imagination. Part of what we want out of a fully realized robotic personality is for it to display something like artificial phronesis, and none of them come close to having those skills. Rather, they are usually nothing much more than a talking search engine; we have only added a robotic interface to common computing technologies. Still, "social robotics is forecasted to expand to more than half a billion dollars by 2023, driven largely by the growing demands of the aging-in-place market that is expected to reach 98 million people in the USA by 2060" (Mitchell 2018).

To succeed, this rapidly growing market is going to have to deliver on its promises of true, friendly, and trustworthy robot companions. A major piece of that puzzle will be solving the problem of programming AP capabilities into those machines. We will now try to rein in the potential misconceptions that can occur when trying to understand what AP is and is not.

In the sections that follow we will find that AP claims that phronesis, or practical wisdom, plays a primary role in high-level moral reasoning; furthermore, AP raises the question of whether or not a functional equivalent to phronesis is something that can be programmed into machines. We will find that the theory is agnostic on the eventuality of machines ever achieving this ability, but that it does claim that achieving AP is necessary for machines to be human-equivalent moral agents. Without AP, artificial agents will have limited abilities to reason ethically (socially constructed behaviors directed toward beneficent outcomes) and no capacity to reason morally (psychologically motivated behaviors for beneficent actions and accurate moral judgment of the observed actions of others). AP is not an attempt to fully describe, or recreate in machines, the phronesis described in classical texts. AP is not attempting to derive a full account of phronesis in humans at either the theoretical or neurological level. AP is not a claim that machines can become perfect moral agents. Instead, AP is an attempt to describe an intentionally designed computational system that interacts ethically with human and artificial agents, even when those situations might be novel and require creative solutions. AP is to be achieved across multiple modalities and most likely in an evolutionary programming or machine-learning fashion. AP acknowledges that machines may only be able to simulate ethical judgment for some time and that the danger of creating a seemingly ethical simulacrum is ever present. This means that AP sets a very high bar against which to judge machine ethical reasoning and behavior. It is an ultimate goal, but real systems will fall far short of this goal for the foreseeable future.

Phronesis: Artificial and Otherwise

There are three levels of artificial agents I want to discuss here, which are classified by their ethical abilities, or lack thereof: ethical impact agents (EIAs), artificial ethical agents (AEAs), and artificial moral agents (AMAs).[1]

EIAs need no explicit claims to phronesis. These systems are notable only in that their operations exhibit some autonomy and have the ability to impact human agents in ways that spur ethical concern. For instance, an autonomous car that, during the course of its autonomous driving operation, hits and kills a pedestrian in an accident (such as the accident that happened on March 18, 2018, in Tempe, Arizona, involving a pedestrian and a self-driving Uber car) does so with no social awareness or any understanding of the practical implications of what has happened. The system's actions are produced by its sensors and actuators working in accordance with a program, but it has no conscious experience of the situation at hand, suffers no moral remorse after the accident, and has no emotional reaction while the event is unfolding. While we might ask plenty of questions about the safety of the autonomous car, no one blames the car itself as a moral agent. That moral condemnation and all questions of legal responsibility are reserved for the emergency backup driver present in the car, the victim, and/or the company that built the car and is testing it on public roads. Thus, there is no particular need to apply AP design considerations when dealing with questions on how to ethically design and deploy EIAs.

The next level of artificial agent is the AEA. AEAs differ from EIAs only in that they have explicit ethical considerations programmed into their operation, either directly or through some sort of machine-learning method. Building on our earlier example, an autonomous car that was programmed to take into account some ethical calculus of value when deciding whether to subject occupants of another vehicle or its own occupants to increased risk during an otherwise unavoidable crash would be an AEA. At first look one might think that an AEA muddies the waters a bit and that the machine itself might deserve moral blame or legal responsibility, but that is just a trick of the light. The moral blame and responsibility for any adverse consequences are still fully borne by the human agents who built, deployed, licensed, and operated the vehicle. The key is that the AEA has not chosen its own ethical standards; instead they were chosen (or, in the case of machine learning, at least accepted) by human agents, who therefore assume any blame or responsibility for any ethical decisions made by the system they designed and/or deployed. The machine is not consciously employing human-like phronetic abilities of any kind to analyze the events that occur based on its operations. It may, however, be deploying early versions of AP, such that it might look to an outside party as if those events were the result of a conscious ethical choice.

It is only when we get to the level of the AMA that conscious phronetic capabilities may play an important role.[2] An AMA would have the ability to choose ethical behaviors that are appropriate to the situation at hand in a way that exhibits a form of practical reasoning similar to what can be seen in the competent ethical reasoning found in most human agents. This means that the system either is a conscious moral agent or is functionally equivalent to one. Another way to say this is to claim that the system displays fully functioning AP. Of course, this concept needs a lot more explanation, and that is what the rest of this chapter will discuss.

While we have been using this term regularly, we must admit that phronesis is a term that many are not familiar with outside of philosophy, and the word can seem a little off-putting. However, if one's goal is to create AI and robotic agents that have the capacity to reason intelligently about ethical situations, then understanding this technical term will reward those who try, given that it is so relevant to the understanding of ethics.

Phronesis has an ancient pedigree and has come down to us largely through the tradition of virtue ethics as the skill of being able to "live well" (Aristotle 1985). The designers of intelligent systems do not necessarily need to become experts on the field of virtue ethics, but it has been shown that some familiarity with the core concepts can enhance the design process both from the level of the designer herself and the ethical reasoning system being designed (Vallor 2016).

Briefly stated, phronesis refers to the practical wisdom that a conscious moral agent uses when she is confronted with a difficult moral or ethical problem and attempts to overcome these difficulties in an intelligent manner. Given that ethical problems are always novel, no set of preconfigured answers will suffice to solve the problem, which means that learning and creativity are the hallmarks of a phronetic agent. "There is no general rule/procedure/algorithm for discerning which values, principles, norms, approaches apply; rather, these must be discerned and judged to be relevant in the first place, before we can proceed to any inferences/conclusions about what to do" (Ess 2009, 25).

AEAs might be successfully designed taking one or more ethical schools of thought into account. One might design an autonomous system that makes ethical decisions based largely on applied utilitarian or Kantian calculations or rules, models of human moral psychology, human religious traditions, or even on the three laws of robotics developed by Isaac Asimov. While one could make AEAs using any of these methods that might be useful in certain circumstances, they will all fall far short of an AMA with artificial phronesis. "For the virtue of practical wisdom or phronēsis encompasses considerations of universal rationality as well as considerations of an irreducibly contextual, embodied, relational, and emotional nature—considerations that Kant and others have erroneously regarded as irrelevant to morality" (Vallor 2016, 24–25).

It is understandable that systems designers may want to ignore ethical reasoning entirely by endeavoring to avoid the construction of EIAs. It would indeed be safest to build autonomous systems that had no conceivable ethical impact on human agents. But the range of applications for this kind of agent is small, maybe just limited to space exploration. This will require that the slightly more adventurous designers who want to enter more lucrative fields, such as social robotics, will need to attempt to apply the more computationally tractable rule-based ethical systems that could result in useful AEAs. Why get involved with AMAs that require something like artificial phronesis to work correctly? To succeed at that may require solving problems in artificial consciousness, artificial emotion, machine embodiment, et cetera, all of which may be computationally intractable. Let's look at what might be the reward for pursuing the more difficult problem of building AMAs with artificial phronesis.

Artificial Phronesis: A Manifesto

Artificial phronesis is the claim that phronesis, or practical wisdom, plays a primary role in high-level moral reasoning; furthermore, it asks the question of whether or not a functional equivalent to phronesis is something that can be programmed into machines.

If we want AI systems to have the capacity to reason about ethical problems in a way that is functionally equivalent to what competent humans do, then we will need to create machines that display phronesis or practical wisdom in their interactions with human agents. This means that AP is one of the highest goals that AI ethics might achieve. Furthermore, this will not be a trivial problem: since not all human agents are skilled at reasoning phronetically, we are asking a lot from our machines if we try to program this skill into them. On top of this, the most difficult problem is that achieving AP may first require that the problem of artificial consciousness be solved, given that phronesis seems to require conscious deliberation and action to be done correctly. Even so, the achievement of AP is necessary, since moral and ethical competence is required in order to develop ethical trust between the users of AI or robotics systems and the systems themselves. Achieving this will make for a future that humans can be comfortable inhabiting, without feeling oppressed by the decisions made by autonomous systems that may impact their lives.

AP is a necessary capacity for creating AMAs; however, the theory is agnostic on the eventuality of machines ever achieving this ability. It does claim, though, that achieving AP is necessary for machines to be human-equivalent moral agents.

AP is influenced by works in the classical ethics tradition, but its sources are not limited only to these. AP is not an attempt to fully describe phronesis as it is described in classical ethics. AP is not attempting to derive a full account of phronesis in humans at either the theoretical or neurological level. However, any advances in this area would be welcome help.

AP is not a claim that machines can become perfect moral agents. Moral perfection is not possible for any moral agent in the first place. Instead, AP is an attempt to describe an intentionally designed computational system that interacts ethically with other human and artificial agents even in novel situations that require creative solutions.

AP is to be achieved across multiple modalities and most likely in an evolutionary machine-learning fashion. AP acknowledges that for quite some time machines may only be able to simulate ethical judgment, and that the danger of creating a seemingly ethical simulacrum is ever present.

This means that AP sets a very high bar against which to judge machine ethical reasoning and behavior. It is an ultimate goal, but real systems will fall far short of this objective for the foreseeable future.

Artificial Phronesis and Human Character

While we are thinking about why we might need something like AP in order to build AEAs and AMAs, it is important to consider what the success of doing so might mean for the development of human phronesis in pursuit of human virtue.

One of the key components of virtue ethics is that in this theory the virtuous agent is always assumed to be learning, growing, and becoming more skilled as a moral agent. Often this system of ethics is seen to be a competitor to or at odds with the other two popular traditions in Western ethics, consequentialist and deontological ethics. In broad terms, a consequentialist ethical framework is about developing more or less complex cost-benefit analysis tools that seek to spread social benefit widely. Deontological systems of ethics are attempts to describe the core set of rules or maxims upon which all specific queries regarding ethical action can be induced or deduced. Each of these systems has significant strengths but also well-known weaknesses and susceptibilities. None of the three is sufficient on its own. Here it is argued that we should look at all three (and any other contenders that the reader may champion) as tools in our ethics toolbox that can be brought out as the situation requires. In the context of computational ethics—or machine ethics, as it is sometimes referred to—there are many successful attempts to devise systems that can reason through simple problems using versions of consequentialist or deontological ethical systems (Wallach and Allen 2009; Anderson and Anderson 2011). As these systems become more complex, they will inevitably run into problems if they try to apply just one ethical theory in a one-size-fits-all manner. What is needed is the same skill human agents have in being able to choose among options and learn when certain ethical problem-solving tools are no longer appropriate. This is where virtue ethics shows its primary value: it helps us master the skill of learning what is best in a given situation and discovering what might be best in a novel situation. Virtue is a kind of fluency with ethical tools. Just as one may be fluent in a language without knowing every word in the dictionary or even all the formal rules of grammar, one can be fluent in ethical actions without knowing every conceivable ethical act or maxim of actions. In fact, it is this creativity that allows for the discovery of new ways of working with ethics and the finding of new tools that are optimized for various situations.

To push the metaphor a bit further, in the same way that no one learns a natural language on their own, They need teachers and others to have conversations with, so too virtuous character needs to be learned by observing the behavior of exemplars. Aristotle calls such a person a *phronemon* (Aristotle 1985). Phronemons provide examples for the neophyte virtuous agent to learn from, first by copied behavior, then more indirectly as a target toward which novel actions are referenced but not explicitly copied from.

What happens when technologies become the exemplars from which behavior is copied by human agents? We might find ourselves in a near future where children interact with smart toys and dolls or smart speakers like the Amazon Alexa that act as if they were fellow agents or even close personal friends. Already children are active users of all sorts of social media technologies. They are naturally inclined to be keenly interested in learning, and often it is machines that have the infinite patience to entertain them when parents are too tired or otherwise distracted.

In this kind of world, we had better have done a good job of constructing machines with beneficial AP, because that AP will be reflected back as the phronemon from which the coming generations will learn to behave.

One quick example might be how robotics could adversely affect the development of the virtues that revolve around the act of trust between humans.[3] Being trustworthy is a skill that is vital for the development of a virtuous character. It is not something that comes naturally, though, and as we develop, we learn the hard way whom we can trust and with whom we should be trustworthy. We do this through the trials of our own interactions with people and through watching others who display skill and mastery at these tasks. Over time (we hope), we develop this virtue, though we can always learn to be better. In the foreseeable future, much of our interactions with the world will be mediated through smart technologies.

Mark Coeckelbergh (2012) has argued that there are two ways of looking at trust and robotics. One is a "contractarian" or more individualistic manner of trust; here we trust the robot not as a fellow agent in the world but as a system that might have certain types of power over us that we have to accept and which we believe will work out to our benefit. This kind of trust is more like an assumption of reliability or safety. His second way of looking at robotic trust is what he calls a "phenomenological" or social level of trust. This is a social world where human agents enter into full social networks that include trusting behaviors between their members. Coeckelbergh does not think that robots will ever be able to enter into this second level of trusting relationships. But he thinks they can engage with us in quasi-trusting relations:

> They may nevertheless contribute to the establishment of "virtual trust" or "quasi-trust" in so far that they *appear* as quasi-others or social others, as language users, and as free agents. We trust robots if they appear trustworthy and they appear trustworthy if they are good players in the social game. (Coeckelbergh 2012, 57)

It follows then that humans who spend most of their time with social robots of one sort or another would have much experience in the contractarian manner of trust but would develop little capacity to trust each other, perhaps in a way similar to the famous short story written by E. M. Forster in 1928 in which he imagined a world where all the human inhabitants lived relatively isolated existences underground where autonomous machines met their every need, but when the machines developed a problem, the humans were too dissociated to actually do anything to stop the destruction of their world (Forster [1928] 1970). It would be a while before anything that drastic happened, but we would see a change in the capacity of people to develop true social-level trusting behavior and the recent large scale COVID related stay at home orders gives us all a small taste of what technological social isolation might be like. We are already seeing these changes in our ability to develop virtues given only simple social networking technologies (Vallor 2016); one can only imagine what will happen with the internet comes right into your living space with you through social robotics technologies. You will replace the many fake human friends you have online with fake robotic friends interacting with you in person.

An additional concern would be the manipulation of trust. When it comes to affective robotic systems that are designed to engage in simulated social relations with their users, we have a perfect opportunity to potentially manipulate users into trusting machines. As the philosopher and roboticist Matthias Scheutz puts it:

> The rule-governed mobility of social robots allows for, and ultimately prompts, humans to ascribe intentions to social robots in order to be able to make sense of their behaviors (e.g., the robot did not clean in the corner because it thought it could not get there). The claim is that the autonomy of social robots is among the critical properties that cause people to view robots differently from other artifacts such as computers or cars. (Scheutz 2012, 207)

This could be mitigated by moving more quickly to machines that had more fully functioning AP capabilities. But it is more likely that we will spend a lot of time in a world dominated by EIAs and AEAs, perhaps for centuries, before we could create full AMAs. This means that we should be very careful with how we design EIAs and AEAs so that we do not fall into the traps that we have outlined here.

Conclusion

Here we have defined artificial phronesis and stated more clearly what this capacity is and is not. It is an attempt to help us build a world not only where our interactions with machines are ethical but also where we can express the best of our human character—one where we can constantly learn to be better and our machines reflect that desire back to us while providing an environment in which we can flourish. We have also learned that this task is not easy and that we are only in the very beginning steps of being able to conceptualize what AP could be, much less being able to effectively program it into machines we are building today.

This means that we have to be very careful about how we introduce social robotics into our lives over the next few decades, since these machines might produce an environment in which the development of human character is reduced.

Notes

* This chapter is dedicated to my family. May the technologies around them help them grow and flourish. I would also like to acknowledge the help of commentators and editors that have helped strengthen this work.
1. This section is adapted from Sullins 2020.
2. Of course, consciousness is a concept that is famously imprecise and difficult to argue for or against. It is not important here if the artificial system has real consciousness or

a very good functional equivalent, but it needs one or the other of these things to be an AMA. See Dietrich et al (2021) for a full discussion of this issue in artificial intelligence technologies.
3. For a full discussion of this topic, see - Sullins, 2020.

References

Anderson, Michael, and Susan Leigh Anderson, eds. 2011. *Machine Ethics*. Cambridge: Cambridge University Press.

Aristotle. 1985. *Nicomachean Ethics*. Edited by T. Irwin. Indianapolis, IN: Hackett Press.

Breazeal, Cynthia. 2002. *Designing Sociable Robotics*. Cambridge, MA: MIT Press.

Coeckelbergh, Mark. 2012. "Can We Trust Robots?" *Ethics and Information Technology* 14: 53–60.

Dietrich, Eric, Christopher Fields, John P. Sullins, Bram von Heuveln, and Robin Zebrowski. 2021. *Great Philosophical Objections to Artificial Intelligence: The History and Legacy of the AI Wars*. London: Bloomsbury.

Ess, Charles. 2009. *Digital Media Ethics*. Cambridge: Polity Press.

Forster, E. M. (1928) 2011. *The Machine Stops*. Penguin classics .

Mitchell, Oliver. 2018. "Jibo Social Robot: Where Things Went Wrong." The Robot Report, June 28. https://www.therobotreport.com/jibo-social-robot-analyzing-what-went-wrong/.

Scheutz, Matthias. 2012. "The Inherent Dangers of Unidirectional Emotional Bonds Between Humans and Soc1ial Robots." In *Robot Ethics: The Ethical and Social Implications of Robots*, edited by Patrick Lin, Keith Abney, and George A. Bekey, 205–22. Cambridge, MA: MIT Press.

Sullins, John P. 2020. "Trust and Robotics." In *Routledge Handbook on the Philosophy of Trust*, edited by Judith Simon, TK–TK. New York: Routledge, pp. 313–326.

Vallor, Shannon. 2016. *Technology and the Virtues: A Philosophical Guide to a Future Worth Wanting*. New York: Oxford University Press.

Wallach, Wendell, and Colin Allen. 2009. *Moral Machines: Teaching Robots Right and Wrong*. Oxford: Oxford University Press.

PART III

VIRTUES AND EPISTEMOLOGY

8

Epistemology, Philosophy of Science, and Virtue

Emanuele Ratti

Introduction

In this chapter, I offer an overview of how virtue-talk and epistemology have been connected, in order to shed light on the epistemic aspects of science. In the epistemology of science, the word *virtue* has referred to two different concepts, each attached to a distinct debate. First, virtue can be understood as excellence, where excellence is a quality of a model, a theory, or a hypothesis. Second, virtue can be understood more narrowly as a stable character trait and/or disposition of scientists themselves. The first meaning is connected to the long-standing debate on the qualities that make a scientific theory a good scientific theory. The second meaning is connected to a much more recent conversation exploring the connections between virtue epistemology and philosophy of science.

Virtues as Excellences of Scientific Products

How does a scientist choose between competing theories? In philosophy of science, this is called the problem of *theory choice*.

According to traditional philosophy of science, there are procedures for choosing among competing scientific theories (see, for instance, Popper 2002; Hempel 1966; Salmon 1967). Ideally, theory choice would be determined by certain criteria that scientific theories should meet in order to be considered good scientific theories. These characteristics have been often called *virtues of theories*, because they are excellences of theories.

Traditional virtues of theories include:

1. Accuracy—how accurate a theory's predictions are
2. Internal coherence—whether a theory hangs together properly, without logical inconsistencies

Emanuele Ratti, *Epistemology, Philosophy of Science, and Virtue* In: *Science, Technology, and Virtues*. Edited by: Emanuele Ratti and Thomas A. Stapleford, Oxford University Press. © Oxford University Press 2021. DOI: 10.1093/oso/9780190081713.003.0009

3. External consistency—whether a theory is consistent with other relevant theories
4. Unifying power—the ability to bring together multiple areas of inquiry
5. Fertility—the ability of a theory to make novel predictions not previously part of the original agenda

Sometimes we refer to these virtues as something that scientists value; hence we might call "unifying power" not just a virtue but also a value. The received view (see, for instance, McMullin 1983 and Longino 1990 for useful introductions) states that virtues of theories are (1) epistemic (i.e., they are truth-conducive) and (2) embedded in a rule-functioning framework (i.e., scientific reasoning deals only with rule-governed type of inference and has no place for any other kind of judgment). However, both of these aspects of the received view have been heavily criticized.

Rules and Virtues

A quick look into the history of science shows how the notion that virtues of theories are embedded in a rule-functioning framework is largely a misleading idealization. Thomas Kuhn in his famous essay "Objectivity, Value Judgement and Theory Choice" (1977) provocatively advances the idea that virtues of theories work as values, in the sense that decisions about theory choice are not akin to proofs. Kuhn identifies five virtues of theories—accuracy (empirical adequacy), internal and external consistency, scope, simplicity, and fruitfulness) — as core criteria for theory choice. If theory choice were a matter of proof and algorithms, one would be able to apply these criteria without difficulty. However, Kuhn argues that there are several problems that make theory choice a serious challenge. First, the criteria are imprecise, in the sense that individuals usually do not agree on how these should be applied to concrete cases. Next, these criteria are repeatedly in conflict with one another—as, for example, is often the case with accuracy and scope (1977, 357). Kuhn analyzes some examples in the history of science where conflicts between these criteria render theory choice far from a clear-cut process. Copernicus's system was not obviously more accurate than Ptolemy's until Kepler drastically revised it later—forty years after Copernicus died. Therefore, Kuhn says, there must have been other reasons to provisionally choose the heliocentric astronomy. Another example is the phlogiston theory, which matched experience in some areas better than oxygen theory, which in turn worked better in other areas of experience. How then do we decide which area of experience is more important for a criterion such as empirical adequacy? Kuhn insists that "the criteria of theory choice with which I began function not as rules, which determine choice, but as values which influence it" (1977, 362). In other words, the five criteria mentioned by Kuhn cannot dictate theory choice as rules, but rather function as values (in the sense of maxims or norms). There are important topics in philosophy of science that are connected to theory choice. Two of these are underdetermination and scientific realism, which are explored with respect to theoretical virtues by Dana Tulodziecki in Chapter 11 in this volume.

McMullin (1983) develops some of these themes. First, he distinguishes *evaluating* and *valuing*. When we evaluate a hypothesis, we measure the extent to which a hypothesis satisfies certain criteria—for example, whether it is empirically accurate, if it is consistent with the corpus of certain disciplines, and so on. By contrast, when we value, we do not try to measure how well a hypothesis meets specific criteria; rather, we discuss the criteria themselves. These criteria are the virtues of theories, which McMullin calls *epistemic values*.

Virtues of theories are epistemic values in the sense of epistemic desiderata. A value is considered epistemic when it is likely to promote those characteristics that make scientific knowledge the type of knowledge usually seen as "the most secure knowledge available to us of the world we seek to understand" (McMullin 1983, 18). It is a desideratum because it is something one believes will help to achieve that kind of knowledge if adequately pursued. McMullin aims to show that theory choice is a procedure much closer to value judgment than to some rule-governed type of inference. By this he means that value judgment is not an unambiguous procedure to determine which choice is the best, but rather a *propensity*—the origin of which is difficult to identify or to track down—to consider a characteristic of an entity to be desirable for an entity of that kind. The judgment involves not the evaluation of how much a particular entity epitomizes that value but, mainly, how much we value that characteristic in that particular entity. For instance, how do we choose to endorse efficiency, fruitfulness, or simplicity? There is no rule for that. Some important questions about science can only be answered by referring to a value judgment of some sort. The fact that theory choice is influenced by value judgments also explains why never-ending controversies are plentiful in science. If there were a methodology or strict rules for choosing among competing theories (i.e., for evaluating theories), then controversies would be easy to resolve.

Feminist philosophers of science such as Helen Longino (1990, 1996) have enriched the catalogue of values important to theory choice. In particular, it has been convincingly shown that typical criteria mentioned by Kuhn are but a small part of the values that one might consider in evaluating competing theories; feminist studies suggest the existence of many other values.

Virtues of Theories and Truth

Let me now turn to the claim that virtues of theories are epistemic. As previously noted, describing something as "epistemic" means that it is *truth-conducive*, in the sense that it directs our inquiry toward truth. In McMullin's view, "truth" is a regulative ideal, a transcendental value of science itself. From this standpoint, McMullin draws the important distinction between *epistemic* and *non-epistemic* values (with examples of the latter being moral, political, and social values). If science is a truth-seeking practice, and if the only values that set science toward truth are epistemic, then the influence of non-epistemic values should be avoided.

However, the epistemic nature of traditional values as truth-conducive has been questioned. In particular, Laudan (2004) argues that what are often called epistemic values in science, such as scope, generality, and the like, are not really epistemic (see also Douglas 2009, ch. 5). Laudan disagrees that features associated with acceptable theories (such as explaining known facts in a domain, explaining different facts, and so on) are associated with epistemology in the sense of being truth-conducive. Rather, Laudan calls such features "cognitive values" (1984, 2004) because they refer to scientists' expectations about good theories that are not related to concerns about veracity. Like Van Fraassen (1980), Laudan thinks that a theory does not have to be true to be good. Nonetheless, cognitive values "are constitutive of science in the sense that we cannot conceive of a functioning science without them, even though they fail to be intelligible in terms of the classical theory of knowledge" (Laudan 2004, 19). Laudan makes an interesting argument by considering Kitcher's "explanatory unification" (1993). Every unifying theory T, Laudan says, must clearly entail non-unifying counterparts $T_1, T_2, \ldots, T_n$. All these non-unifying counterparts must be true if T is true. If T and its counterparts are all true but scientists regard T as better than its counterparts, this is because T has a non-epistemic virtue (the virtue of explanatory unification) that its counterparts do not possess, though the counterparts are not false or less true (whatever this means). Ergo, explanatory unification (like other traditional desiderata for theories) is not really epistemic.

Douglas (2009) makes finer-grained distinctions. She defines cognitive values as "those aspects of scientific work that help one think through the evidential and inferential aspects of one's theories and data" (2009, 93). These values stimulate the possibility of scientific work in the immediate future. But she does not identify cognitive values with what were once called epistemic values. While Laudan seems to hold an eliminativist position with respect to epistemic values, Douglas thinks that there are a few of these epistemic values. However, they are not what one might think they are. Values such as internal consistency and predictive competency are conceived as epistemic in the sense that if a theory lacks them, then the theory has serious epistemic problems. For example, if a theory is not internally consistent, then it is surely false. So epistemic values operate in a negative way: "without meeting these criteria, one does not have acceptable science" (Douglas 2009, 94).

But concepts of "virtue" may refer not just to the products of science but also to the scientists themselves. In this case the excellence is about one or more features that a scientist possesses, features that make her a good scientist. While the literature in philosophy is rather scarce on this topic, recent connections between virtue epistemology and philosophy of science have the potential to tackle this issue. The next section provides an overview of these connections.

Virtues, Virtue Epistemology, and Science

Throughout its history, analytic philosophy of science has been theory/model/explanation-centered. There is of course a widely shared awareness that a scientist's intentions or specific perspective (see, for instance, Van Fraassen 2008; Giere 2006) are likely to play an important role in the practice of science. However, this insight has rarely been deepened, or at least not deepened in the way that one may expect. As a general trend, philosophers of science have focused on the products and methodologies of science rather than on the scientist, her virtues, and the epistemic consequences (if any) of cultivating certain habits instead of others. Despite the impressive plethora of studies on the role of non-epistemic values in undermining the ideal of value-free science, most contemporary philosophy of science seems to favor the ideal of the selfless scientist, or at least a scientist whose self—as a source of biases—is limited in important ways (Laura Ruetsche makes a similar point at the beginning of Chapter 9 in this volume). To use another formulation, a substantial part of philosophy of science has focused mainly on the medium used to formulate scientific knowledge (theories, models, explanations, etc.) and the relation of these outputs to the natural world. What has been neglected in the philosophical studies of science is the *scientist*. The so-called practice turn in philosophy of science (Soler et al. 2014), even if it is focused on the material and social conditions underlying the production of knowledge, tends to consider the scientist and the important contribution of her character traits a black box. There is an important exception to this trend: some philosophers have started to draw an interesting comparison between a field called virtue epistemology and philosophy of science.

Virtue Epistemology

Traditional analytic epistemology is understood as "belief-based" epistemology (Battaly 2008), in the sense that beliefs are the primary objects of epistemic evaluation. This means that other important concepts in analytic epistemology (knowledge, justification, etc.) are built on the very concept of "belief." In other words, "belief" is the most fundamental concept of traditional epistemology, and knowledge is defined in terms of the relation between beliefs and evidence (whether the belief is true), as well as in terms of the way beliefs are formed (whether the belief is justified). Virtually all analytic epistemologists have thought that the right definition of knowledge must be in terms of true and justified beliefs. However, there are different conceptions of justification, and there are at least six necessary conditions attached to the concept of justification that are all belief-based (see Alston 1993). It is possible to trace an interesting parallel between the belief-based nature of analytic epistemology and general philosophy of science by relying on what I said earlier about philosophy of science being focused on the medium used to formulate scientific knowledge (such as theories, models, explanations, etc.).

Virtue epistemology is an approach in analytic epistemology that stresses the importance of agents/knowers rather than beliefs. This is not to say that beliefs are unimportant in virtue epistemology. On the contrary, beliefs remain central, but the evaluation of the agent/knower is more fundamental than the concept of "belief" itself, which is defined mostly in terms of the features of the knower/agent. In other words, in virtue epistemology the epistemic concerns shift from the abstract structure of beliefs to the intellectual structure of the knower/agent. We no longer define key concepts such as "knowledge" in terms of beliefs, since belief is not considered as fundamental and foundational as analytic epistemologists think.

Which features of an agent are we referring to? As the label *virtue epistemology* suggests, the features of the agent through which the concerns of analytic epistemology are rechanneled are called *virtues*. But, as we know very well, the term *virtue* encompasses different traditions and is used with different meanings, even within the virtue epistemology debates. What the different strands of virtue epistemology have in common (even if they use the term *virtue* in different ways) is the fact that they think that the characteristics of the agent/knower (be they faculties, skills, or virtues in the Aristotelian sense) play a role (sometimes indispensable, at other times only important) in evaluating whether a belief constitutes knowledge. In other words, knowledge would be a justified true belief arising from (intellectual) virtue.

According to several review articles (Battaly 2008; Greco and Turri 2011), the literature on virtue epistemology might be decomposed in light of two fundamental dichotomies.[1] The first is between theorists and anti-theorists. Theorists support the possibility of a full-fledged theory connecting virtues and knowledge (Sosa 1991; Zagzebski 1996). On the other hand, anti-theorists argue that while virtues are important to knowledge, systematic connections between the two are not possible (Kvanvig 1992; Roberts and Wood 2007). The second dichotomy is between reliabilists (Sosa 1991; Greco 2000, 2010) and responsibilists (Zagzebski 1996; Montmarquet 1993; Roberts and Wood 2007; Baehr 2011). While almost all virtue epistemologists agree that virtues are excellences and that intellectual virtues are cognitive excellences, how the term *cognitive* should be understood is controversial. In other words, there is a disagreement on the nature of virtues. This latter dichotomy is especially relevant to the philosophy of science, and I will explore it in more detail.

So-called reliabilists (e.g., Sosa, Greco) think that an "intellectual virtue is a quality bound to help maximize one's surplus of truth over error" (Sosa 1991, 225). Sosa first mentions rationalist intuition and deduction as exemplary excellences, then further develops his notion of "virtue." By asking "what other faculties might one admit" (Sosa 1991, 225), Sosa seems to be thinking about virtues in terms of faculties, and if intellectual virtue is a sort of excellence, then intellectual virtues would be well-functioning faculties, including sense perception such as vision. In other words, *virtue* and *faculty* are basically synonyms. This means that virtues are not acquired (as is usually held in the Aristotelian tradition) but might be innate. Sosa (1991, 271) explicitly departs from Aristotle in the way he understands virtues; he opts for a broader sense.

According to reliabilists, intellectually virtuous faculties are the source of true beliefs, and knowledge is a true belief out of intellectual virtue. In other words, knowledge is a true belief produced by a well-functioning and reliable faculty. Sosa, Greco, and the others supporting this class of positions are called reliabilists because of the focus on virtues as reliable faculties.

Responsibilists explicitly draw the concept of "virtue" from the Aristotelian tradition. As Battaly (2008) reports, the ideas of virtue responsibilists (Zagzebski, Montmarquet, Roberts, Wood, etc.) stem from the intuition that an agent is intellectually virtuous because of some active features, including motivations, actions, and habits for which she is ultimately responsible (from this comes the label *responsibilism*). In a very broad sense, intellectual virtues are acquired traits of intellectual action and motivation, and since we are not responsible (or, let's say, we are responsible to a lesser degree) for faculties to be perfected, then most of the virtues cited by reliabilists will not count as virtues for responsibilists.

Zagzebski (1996) was the first to explicitly draw a full-fledged parallel between moral theories and epistemology. She began by noticing that analytic epistemology is constructed along the same lines as modern moral theories. In her opinion, for virtually all epistemic theories it is possible to draw a parallel with modern moral theories. Modern moral theories take the individual act as the fundamental unit of analysis; analytic epistemology takes the belief to be that fundamental unit. There is not, in Zagzebski's opinion, attention paid to the agent. To avoid many flaws related to this commonality between modern moral theories and analytic epistemology, she supports the idea that epistemology should be framed in terms of virtue theories rather than other types of moral theories. By focusing explicitly on the Aristotelian tradition, Zagzebski defines a virtue (both intellectual and moral) as "a deep and enduring acquired excellence of a person, involving a characteristic motivation to produce a certain desired end and reliable success in bringing about that end" (1996, 137).[2] Motivation and success are two hallmarks of virtues: they are motivated by the desire for truth (the motivation component of a virtue), and they are truth-conducive (the success component of a virtue). This is a very demanding notion of virtue, and other responsibilists, while endorsing an Aristotelian perspective, elaborate notions of virtues with fewer requirements. Roberts and Wood, for instance, think of intellectual virtues as "acquired bases of excellent intellectual functioning" (2007, 60), where such traits "make a person excellent *as a human being*" (2007, 68), but they deny that each intellectual virtue has its own motivation and success components. Baehr defines intellectual virtues as "'as personal intellectual excellences' or as traits that contribute to their possessor's 'personal intellectual worth'" (2011, 89), but he downplays the importance of the success component.

Virtue Epistemology and Science

Recent attention to virtue epistemology in philosophy of science can be traced in part to David Stump's (2007) revival of Pierre Duhem's concept of "good sense." Famously, Duhem (1954) argues that when empirical evidence is underdetermined, scientists do not choose theories in light of well-specified rules, in the sense that "there is no formal method by which to make a decision" (Stump 2007, 155). According to Stump's reconstruction of Duhem's argument, "we have an intuitive reasoning ability, which Duhem terms 'good sense,' that allows scientists, like judges in a legal setting, to be able to weigh evidence and to be fair and impartial" (2007, 149–50). In particular, Duhem says, "the day arrives when good sense comes out so clearly in favor of one of the two sides that the other side gives up the struggle even though pure logic would not forbid its continuation" (1954, 218). Stump emphasizes that "good sense"—a puzzling concept for philosophers of science—should be understood first as an ethical term. But if "good sense" is decisive for theory choice, we should admit that "intellectual and moral virtues [such as 'good sense'] of the scientist determine scientific knowledge" (2007, 150). This means that such virtues are not just ethical concepts but have an epistemological dimension as well. According to Stump, this interaction of the moral and the epistemic can be properly understood in light of virtue epistemology, especially responsibilists' views. These considerations would make Duhem a sort of virtue epistemologist *ante litteram* because in a sense they would make (scientific) knowledge dependent on the agent's personal traits.

This interpretation of Duhem's epistemology has been criticized. For instance, Ivanova (2010) argues that the aims of Duhem and the aims of virtue epistemologists are rather different. In particular, Duhem denies the possibility of a perfect and ultimate true scientific theory, while virtue epistemologists aspire to something like that. Kidd (2011) argues that Ivanova's concerns apply only to reliabilists' virtue epistemology, while responsibilists have a weaker epistemic aim. Moreover, Kidd goes on, Duhem's view of the impossibility of ever reaching a true scientific theory may be interpreted as responsibilists' virtue of humility.

While this debate at first was mainly focused on Duhem's interpretation (Ivanova 2011; Fairweather 2012; Ivanova & Paternotte 2013), recently it has been concerned with more general questions about the role of virtues in the construction of scientific knowledge. In particular, Paternotte and Ivanova (2017) scrutinize the topic of virtues and science (in particular whether virtues are beneficial to theory choice) by using tools from history of science and formal epistemology, concluding that while virtues are not necessary for theory choice, sometimes they can work as catalysts for theory convergence.

An attempt to make a connection between virtue epistemology and the nature of science that can also work as a partial response to Ivanova and Paternotte's arguments comes from Baehr (2011), a virtue epistemologist. Baehr's strategy is to show how reliabilism can work also for high-grade knowledge, but only when one considers intellectual virtues as character traits in general, which is how responsibilists understand virtues (in this context, "low-grade" knowledge refers to perceptual knowledge such as "I know I am in a room with four walls," while "high-grade knowledge" refers to much more complex knowledge, such as scientific knowledge). Reliabilists think that a personal quality is an intellectual virtue if it plays a critical role in being truth-conducive. Therefore, faculties of various sorts (such as deductive reasoning and vision) function as intellectual virtues. However, this framework works only for low-grade knowledge. When high-grade knowledge is concerned, the role of such faculties is surely required, but it is uncontroversial. On these premises, Baehr argues for a reliabilist perspective on high-grade knowledge strictly connected with a responsibilist notion of virtue as acquired character trait. The general idea advanced by Baehr is that personal agency as it relates to character traits "makes a salient contribution to what we might call the 'evidential situation' of the person in question, meaning that it largely determines either the *content* of the person's evidence or how the person *handles* or *regards* this evidence" (2011, 82). This means that, on the basis of the same data, different character traits (be they virtues or vices) make scientists see different kinds of evidence.

Epistemic Injustice

A topic standing at the intersection of epistemology, philosophy of science, and ethics is *epistemic injustice* (a term coined by Miranda Fricker [2007]). In general, epistemic injustice is when someone is wronged in her capacity as a knower; in other words, it involves "obstacles to activities that are distinctively epistemic" (Grasswick 2017, 314). In particular, she refers to two types of injustices that are genuinely epistemic. First, testimonial injustice is when prejudices or biases cause an agent to assign a lower level of credibility to another agent's word. Second, hermeneutical injustice occurs when one individual's or group's social experience is "obscured from collective understanding" (2007, 158) by a lacuna that may exist for a number of reasons. She distinguishes between incidental and systematic hermeneutical injustice: the former occurs when there is a lacuna in the collective hermeneutical resources, while the latter occurs when the lacuna is there because of biases and prejudices (and hence is persistent). Fricker also associates two virtues that can balance those injustices: testimonial justice and hermeneutical justice. Works on epistemic injustice are interesting because they emphasize that ethical and epistemic questions are strictly connected and can be separated only in the abstract. There is also a growing literature that aims to analyze scientific practices using this notion as a lens (see, for instance, Grasswick 2017 and her analysis of participatory injustice and epistemic trust injustice in science), though several works in feminist philosophy of science can be read in this way as well.

Conclusion

The role of virtue and virtue theories is quite underexplored in studies of epistemology of science. Apart from the debate on Duhem and the article by Paternotte and Ivanova (2017), epistemologists of science does not seem to have realized the importance that virtue theories may have for understanding some epistemic aspects of science. As I mentioned before, even the practice turn in philosophy of science (and science studies in general), while it has explored in detail the contribution of social structures and material conditions to our understanding of the nature of science, has yet to realize the full potential of the role of habits and other characteristics of scientists play in constructing a comprehensive picture of the epistemology of science.

Notes

1. Authors on the opposite side under the lens of one dichotomy might be on the same side in light of the other. Paradigmatic examples in this respect are Sosa and Zagzebski.
2. Actually, Zagzebski thinks that intellectual virtues are a subset of moral virtues.

References

Alston, W. 1993. "Epistemic Desiderata." *Philosophy and Phenomenological Research* 53, no. 3: 527–51.

Baehr, J. 2011. *The Inquiring Mind*. Oxford: Oxford University Press.

Battaly, H. 2008. "Virtue Epistemology." *Philosophy Compass* 3, no. 4: 639–63.

Douglas, H. 2009. *Science, Policy, and the Value-Free Ideal*. Pittsburgh: University of Pittsburgh Press.

Duhem, P. 1954. *The Aim and Structure of Physical Theory*. Princeton: Princeton University Press.

Fairweather, Abrol. 2012. "The Epistemic Value of Good Sense." *Studies in History and Philosophy of Science Part A* 43 (1). Elsevier Ltd: 139–46.

Fricker, M. 2007. *Epistemic Injustice*. Oxford: Oxford University Press.

Giere, R. 2006. *Scientific Perspectivism*. Chicago: University of Chicago Press.

Grasswick, H. 2017. "Epistemic Injustice in Science." In *The Routledge Handbook of Epistemic Injustice*. Routledge. Pp 313–323.

Greco, J. 2000. *Putting Skeptics in Their Place*. Cambridge: Cambridge University Press.

Greco, J. 2010. *Achieving Knowledge*. Cambridge: Cambridge University Press.

Greco, J., and J. Turri. 2011. "Virtue Epistemology." In *Stanford Encyclopedia of Philosophy*.

Hempel, Carl. 1966. *Philosophy of Natural Science*. Upper Saddle River, New Jersey: Prentice-Hall.

Ivanova, M. 2010. "Pierre Duhem's Good Sense as a Guide to Theory Choice." *Studies in History and Philosophy of Science Part A* 41, no. 1: 58–64.

Ivanova, M. 2011. " 'Good Sense' in Context: A Response to Kidd." *Studies in History and Philosophy of Science Part A* 42, no. 4: 610–12. https://doi.org/10.1016/j.shpsa.2011.09.006.

Ivanova, M., and C. Paternotte. 2013. "Theory Choice, Good Sense and Social Consensus." *Erkenntnis* 78, no. 5: 1109–132.

Kidd, I. J. 2011. "Pierre Duhem's Epistemic Aims and the Intellectual Virtue of Humility: A Reply to Ivanova." *Studies in History and Philosophy of Science Part A* 42, no. 1: 185–89.

Kuhn, T. 1977. "Rationality, Value Judgment, and Theory Choice." In *The Essential Tension*, 320–39. Chicago: University of Chicago Press.

Kvanvig, J. 1992. *The Intellectual Virtues and the Life of the Mind*. Savage, MD: Rowman & Littlefield.

Laudan, L. 2004. "The Epistemic, the Cognitive, and the Social." In *Science, Values and Objectivity*, edited by P. Machamer and G. Wolters, TK–TK. Pittsburgh: University of Pittsburgh Press, pp. 14–23.

Longino, H. 1990. *Science as Social Knowledge: Values and Objectivity in Scientific Inquiry*. Princeton: Princeton University Press.

Longino, H. 1996. "Cognitive and Non-Cognitive Values in Science: Rethinking the Dichotomy." In *Feminism, Science, and the Philosophy of Science*, edited by L. H. Nelson and J. Nelson, 39–58. Dordrecht: Kluwer Academic.

McMullin, E. 1983. "Values in Science." *PSA: Proceedings of the Biennial Meeting of the Philosophy of Science Association* 2: 686–709.

Montmarquet, J. 1993. *Epistemic Virtues and Doxastic Responsibility*. Lanham, MD: Rowman and Littlefield.

Paternotte, C., and M. Ivanova. 2017. "Virtues and Vices in Scientific Practice." *Synthese* 194: 1787–807.

Popper, Karl. 2002. *The Logic of Scientific Discovery*. London and New York: Routledge.

Roberts, R., and J. Wood. 2007. *Intellectual Virtues: An Essay in Regulative Epistemology*. Oxford: Oxford University Press.

Salmon, W. 1967. *The Foundations of Scientific Inference*. Pittsburgh: University of Pittsburgh Press.

Soler, L., S. Zwart, M. Lynch, and V. Israel-Jost, eds. 2014. *Science After the Practice Turn in the Philosophy, History and Social Studies of Science*. New York: Routledge.

Sosa, E. 1991. *Knowledge in Perspective*. Cambridge: Cambridge University Press.

Stump, D. J. 2007. "Pierre Duhem's Virtue Epistemology." *Studies in History and Philosophy of Science Part A* 38, no. 1: 149–59.

Van Fraassen, B. 1980. *The Scientific Image*. Oxford: Clarendon Press.

Van Fraassen, B. 2008. *Scientific Representation: Paradoxes of Perspective*. Oxford: Oxford University Press.

Zagzebski, L. 1996. *Virtues of the Mind: An Inquiry into the Nature of Virtue and the Ethical Foundations of Knowledge*. New York: Cambridge University Press.

9

Virtue and Contingent History

Engineering Science*

Laura Ruetsche

Physics as a whole is always in a state of incomplete coordination between extraordinarily diverse pieces of its culture: work, machines,
evidence, and argument. That these messy pieces come together as
much as they do reveals the presence, not of a constricted calculus of
rationality, but of an expanded sense of reason.

—Peter Galison (1997, xxii)

Introduction

In *Bayes or Bust*, John Earman expresses a thought, one that appears to be widely
shared, about what an adequate epistemology of science might look like:[1]

> To my mind, the most interesting aspect [of science] is the epistemic one. I in-
> sist . . . that this aspect be explained in Bayesian terms. This implies that all valid
> rules of scientific inference must be derived from the probability axioms and
> the rule of conditionalization. (Earman 1992, 204)

Earman's thought is that a constricted calculus of rationality—the probability
calculus and the rule of conditionalization—suffices to characterize the epistemic aspect of science, including the phenomena of scientific knowledge and
empirical justification. But what if a rationality bound by a constricted calculus
is a rationality trammeled and inhibited from accomplishing all that it might?
Here I'll offer a picture of a relatively untrammeled variety of rationality—an expanded sense of reason—and the epistemic aims its operation in the sciences
might advance. In developing this picture, I'll rely on a notion of virtue as (some
think) Aristotle conceived it: virtue as a second-nature capacity to see reason,
a capacity inculcated through having lived a particular concrete history, a capacity not reducible to the application of a formal calculus. The approach to an

Laura Ruetsche, *Virtue and Contingent History* In: *Science, Technology, and Virtues*. Edited
by: Emanuele Ratti and Thomas A. Stapleford, Oxford University Press. © Oxford University Press 2021.
DOI: 10.1093/oso/9780190081713.003.0010

expanded sense of scientific reason I sketch here has implications for the social structure of science. I'll try to gesture at some of those implications and some questions they raise.[2]

But I have some obligations to discharge first. My mission is to present *virtue* as a way of thinking and talking about an expanded sense of reason operating in science. Natural questions prompted by this mission include: Expanded *compared to what*? And *why* should this expanded sense of reason matter to *epistemology*? The next two sections will tackle these questions. Then the final section will address the question of *how* to care about the variety of reason I try to describe here. That is, it will consider epistemic engineering consequences of the idea that at least some sensitivities to reason are importantly analogous to the sort of virtue wrought by having lived an appropriate contingent history.

A Constricted Calculus of Rationality

To draw attention to the features of "mainstream" epistemology she takes to be problematic, Sandra Harding (1986) cites Francis Bacon's invocation of "a way of discovering sciences [that] goes far to level men's wits, and leaves but little to individual excellence; because it performs everything by surest rules and demonstrations" (Bacon 1960, I.cxxii). Bacon's invocation suggests a mission for the epistemologist: analyze *without residue* the epistemologically central commodity of the *justification* of belief in terms of these rules and demonstrations. I'll use the label "Traditionalist" for epistemologies that take their mission to be coextensive with the one Bacon describes. Traditional epistemologies of science seek to characterize a constricted calculus of rationality governing empirical justification.

Earman's aspiration to explain the "epistemic aspect" of science by deriving "all valid rules of scientific inference" from "the probability axioms and the rule of conditionalization" expresses a commitment to Traditionalism. The form of Traditionalism Earman espouses is known as *Bayesian subjectivism*. I'll explicate the approach in more detail, in order to articulate some key features of it. By using Bayesian subjectivism as an example, I don't mean to endorse it, nor do I mean it to imply that all Traditionalists, much less all mainstream epistemologists, are Bayesians. Bayesian subjectivism is a good example to use here because it is meant to be both modest and resourceful in acknowledging the sorts of influences on successful science that escape the confines of a constricted calculus of rationality.

A subjectivist interprets probabilities as subjective degrees of belief; roughly, the probability that an agent assigns a hypothesis reflects her level of confidence in the hypothesis and can be gleaned from the betting quotients she'd accept as fair odds for a wager on the hypothesis. If the agent's degrees of belief violate axioms of the probability calculus, she's prone to Dutch books—sets of bets at such odds that she's guaranteed to lose money. So, supposing it's rational to avoid Dutch books, her initial assignment of probabilities is, if rational, constrained by the axioms of the probability calculus. But only by those axioms: subject to the constraint, her initial assignment can embody stubborn prejudice and selfish fantasy.

As the agent makes her way through the world, she gathers evidence. If the agent is rational, she will, obeying the rule of conditionalization, respond to this evidence by revising her degrees of belief in such a way that her new probability function is just her old probability function conditionalized on the novel evidence. Bayes's rule, an axiom of the probability calculus, governs the details of the construction of her new probability function from her old one.

According to the Bayesian subjectivist form of Traditionalism, *good epistemic practice consists in updating one's subjective degrees of belief in response to new evidence by means of the rule of conditionalization and Bayes's rule.*

Bayesians emphasize just how modest a constraint this is. Deductive logic, we tell disappointed introductory students, will not tell us what to believe. It will tell us only, given that we believe some things, what other things we ought not—on pain of contradiction—believe. Just so, none of the Bayesian apparatus dictates our initial degrees of belief or our prior probability assignments. It tells us only how those degrees of belief ought—on pain of susceptibility to unscrupulous bookies—change in light of incoming evidence. What about influences other than a commitment to conform to inductive inference? Bayesian subjectivism leaves them plenty of room to maneuver in the epistemic life of a scientific agent. For instance, such influences could have (just about) everything to do with the assignment of prior probabilities to hypotheses. Degree of belief revision, Bayesians say, must nevertheless proceed according to the rule of conditionalization and Bayes's rule, rules to which agents are beholden regardless of their prior probability assignments, their social stations, their hopes, desires, or plans. Following those rules gives agents epistemic justification, not to hold beliefs outright, but to revise degrees of belief in light of incoming evidence.

One reason to follow valid rules of scientific inference is that we thereby attain "objective" belief, belief conforming to objects (as opposed to subjects) in that it is supported by evidence rather than by parochial or idiosyncratic allegiances. Hence Bayesians' delight with "washing out" results—results showing that, (almost) no matter what initial degrees of belief (prior probabilities) agents bring to inquiry, if those agents observe the same string of evidence and update their degrees of belief using Bayes's rule, they will eventually converge in their probability assignments.[3] Driven by evidence rather than by suspicion, this convergence promises to be objective in a way the initial assignments of personal probabilities are not.

Earman takes epistemic warrant to be governed by "valid rules of scientific inference," which he seeks to codify by appeal to this Bayesian apparatus. I'll call epistemic warrant characterized by adherence to a constricted calculus of rationality, such as Bayesian confirmation theory, *rarefied warrant*. Successfully articulating the rules of rarefied warrant *exhausts* the task of epistemology, Traditionally construed. Excise from science instances of following rules of rarefied warrant and no epistemic residue remains.

Traditionalists needn't remain silent about aspects of science other than its justificatory practices and the comprehensive rules they follow. They just have to deny *epistemic* status to these additional aspects—as Wes Salmon does when discussing N. R. Hanson's logic of discovery (Hanson 1960). Salmon takes Hanson to highlight "plausibility considerations" to the effect that some methods of hypothesis formation are more likely to generate true hypotheses than others. Acknowledging that such plausibility considerations do exist and do influence the practice of science, Salmon nevertheless distinguishes them from the justification achieved by "testing or confirmation," which he identifies as the proper concern of philosophers of science (Salmon 1966, 111–14). Denying *epistemic* interest to plausibility considerations eluding characterization via a calculus of confirmation, Salmon expresses a commitment to Traditionalism.

Enough has been said to distill key features of Traditionalism and the constricted calculus of rationality it pursues: Traditionalism distinguishes the contexts of discovery and justification. It locates the epistemic achievements worth tracking in the latter and devises for their capture a net taking the form of an inductive logic or confirmation theory, a codification of the extent to which evidence (wherever it comes from) supports hypotheses (wherever they come from). Every inference conforming to the rules codified is epistemically warranted, regardless of the particularities of its setting; every attainment of epistemic warrant in the sciences is an instance of an inference conforming to the rules. Every rational agent has what it takes to achieve rarefied warrant—and so *differences* between agents, their contexts, their histories, and so on, *hold no interest* for the epistemologist.

Critics of mainstream epistemology of science seek a "less partial and distorted" (Harding 1991, 1) way to understand scientific knowledge. Some critics express a suspicion of the very idea of what I've been calling rarefied warrant, understood as a species of empirical justification accessible to all and codifiable by a universally applicable calculus of rationality. I am content here to waive that suspicion and agree that Traditionalists set out to characterize an epistemic phenomenon worth our attention. However, Traditionalism can be partial (in the sense of incomplete, if not in the sense of unfair) if responsible scientific practice involves epistemic achievements that do not consist in attaining rarefied warrant. If such epistemic achievements exist, Traditional epistemology is incomplete. It becomes distorted if it appends to its analysis of rarefied warrant the claim that attaining such warrant is the only kind of epistemic achievement there is.

Traditionalism could be missing something. The notion of *virtue* affords a way to characterize what it's missing and how and why it matters that Traditionalism is missing it.

An Expanded Sense of Reason

This section takes on three tasks. The first is to offer an account of Aristotelian virtue that articulates virtue as a model of an epistemic capacity that eludes characterization by means of a constricted calculus of rationality, such as Bayesian subjectivism. The second task is to suggest that virtue-like epistemic capacities do operate in the sciences. If this suggestion is correct, Traditionalism is incomplete. Traditionalism can disarm this charge of incompleteness by contending that the capacities here likened to virtue aren't genuinely *epistemic*. The third task is to make a case that the capacities in question *do* deserve regard as genuinely epistemic—as well as to expose the epistemological presuppositions underlying that case.

I should hasten to emphasize that I don't mean here to take a stand on how to interpret Aristotle's ethics. I mean rather to develop a way of speaking (modeled, it happens, on a way of speaking suggested to some by Aristotle's ethics) that turns out to be apt for probing possible limits of Traditionalism in the epistemology of science.

For Aristotle, man is by nature rational. But, Aristotle tells us in the *Nicomachean Ethics*, he is not by original nature virtuous: "The virtues are engendered in us neither by nor yet in violation of nature; nature gives us the capacity to receive them, and this capacity is brought to maturity by habit" (II.i).[4] The virtuous man has through careful training developed what we might call a *second nature* to respond in the morally appropriate way to his circumstances. The virtue acquired by cultivating the appropriate second nature is not a placeholder for a set of dispositions to deploy some moral calculus: "Still less is exact precision possible in dealing with particular cases of conduct; for these come under no science or professional tradition, but the agents themselves have to consider what is suited to each circumstance on each occasion" (II.ii). Virtue is not encoded in general rules, transparent to our inherent first nature, but rather is a capacity born of contingent histories, inculcated second natures.

Aristotle's virtuous man can act appropriately without following universally transparent principles of right conduct. Does he therefore act without reason?

Only if exercises of reason must take the form of following rules transparent to all. In his 1976 paper "Virtue and Reason," John McDowell attributes to Aristotle an alternative model of rational action. For his Aristotle, the contingently ennatured sensitivity to moral demands that is virtue consists not in following rules but rather in exercising something like "a perceptual capacity," a capacity, as it were, to see reason (McDowell 1976, 332). The exercise of moral reason available to Aristotle's virtuous man is not available to those who, lacking the benefit of a proper upbringing, haven't developed the appropriate second nature, the appropriate sensitivities to patterns of moral salience that the virtuous man has.

With a view toward transferring these notions to the scientific arena, it is worth noting that the position developed so far is consistent with an unabashed moral realism. Such a realism would hold that not every capacity—not even every socially habituated and celebrated capacity—is a virtue. Only the capacities the exercise of which constitutes getting things morally right—where moral rightness does not reduce to some conglomerate of (social) context-sensitive features—are virtues. Such a moral realist can maintain that our access to virtue is socially conditioned without adopting a relativism according to which *what virtue is* is determined by social context. On this view, moral reality is not a creature of context, but the capacity to track it, the capacity I've labeled "virtue," is.

The story thus far: Aristotelian virtue is a sensitivity to reason that is (1) contingently engendered (that is, whether one develops it or not depends on one's social, historical, political, etc. context) and (2) resistant to subsumption under universal rules. McDowell's stated aim in developing his take on Aristotelian virtue is to undermine the idea that "ethics [is] a branch of philosophy related to moral theory . . . rather as the philosophy of science is related to science" (1976, 331)—the idea, that is, that just as it is the task of the philosopher of science to elucidate scientifically justified belief by articulating the "rules of valid scientific inference" to which it conforms, so too is it the task of the ethicist to elucidate right conduct by articulating the "principles of acceptable behavior" to which it conforms (1976, 331). McDowell protests that the analogy banishes virtue, understood as the exercise of a second-nature sensitivity to moral reasons, from the ethicist's landscape. He invokes Aristotle to inspire us to reinstate it.

Suppose that McDowell has persuaded us about ethics and moral theory. We should not therefore lose interest in the analogy

ethics : moral theory :: philosophy of science : science

For now we can ask: should philosophy of science ever stand to science as McDowell would have ethics stand to moral theory? That is, are there epistemic capacities operating in the sciences analogous to the practical rationality that Aristotle's virtuous man, exercising his contingently ennatured perceptual capacity to see aright, deploys? If so, the epistemic achievements attained by exercising such capacities could not be assimilated to (for example) obedience to an inductive logic or confirmation theory, or to any other model that takes epistemic achievement to consist only in adherence to a template provided by universal rules.

Traditional epistemology offers models of scientific warrant that suppose it to conform to such a template. But if second-nature epistemic capacities matter in the sciences, even an adequate account of rarefied epistemic practices falls short of being an adequate epistemology of science.

Something like the suggestion that scientific practice includes, and includes critically, second-nature epistemic achievement has been made by a host of philosophers. The host includes Nancy Cartwright (1987), Ian Hacking (1983), Peter Galison (1997), and (maybe even) Mark Wilson (2018). Members of the host would shift the locus of critical scrutiny from physical theory, conceived as a global, complete, logico-mathematical entity, to the localized, in some sense fragmented, practices and interactions of theorists, applied mathematicians, experimentalists, and their equipment. Traditional epistemology relegates the question of where hypotheses come from to the capricious context of discovery, and relegates the question of where evidence comes from to the machinations of laboratory technicians, so that it might focus on the question of the extent to which evidence (wherever it comes from) warrants hypotheses (wherever they come from). The philosophers I've just mentioned draw our attention to activities that Traditional epistemology of science thereby neglects—activities such as seeing through a microscope or a telescope, using instruments to tell phenomena from artifacts, knowing your recalcitrant laboratory apparatus well enough to keep them working, crafting phenomenological laws and other useful approximations, communicating with representatives of scientific subcultures other than the one in which you were raised, and so on.

Hacking, for example, insists:

> Noting and reporting readings of dials—Oxford philosophy's picture of experiment—is nothing. Another kind of observation is what counts: the uncanny ability to pick out what is odd, wrong, instructive or distorted in the antics of one's equipment. (1983, 230)

"[Such] observation is a skill," he declares. "Some people are better at it than others. You can often improve this skill by training and practice" (1983, 168). Caroline Herschel is a case in point.

> I think that Caroline discovered more comets than any other person in history. She got eight in a single year. Several things helped her do this. She was

indefatigable. . . . She also had a clever astronomer for a brother. . . . When she did find something curious "with the naked eye," she had good telescopes to look more closely. But most important of all, she could recognize a comet at once. Everyone except possibly brother William had to follow the path of the suspected comet before reaching any opinion on its nature. (1983, 180)

Donna Haraway describes another example, one to which we'll return, of evidently acquired abilities to see:

[Irven] DeVore [one of primatologist Sherwood Washburn's male students] *literally saw* a male-centred baboon troupe structure . . . [Phyllis] Jay [one of Washburn's woman students] *explicitly saw* the infant as a key centre of attraction in langur troop structure . . . she *literally, physically saw* what almost could not figure in her major conclusions because another story ordered what counted as ultimate explanation. (1991, 95, 96; emphasis added)

What drops out of what Hacking calls "the Oxford picture" is the fact that to focus exclusively on the inferential significance of an observation—its location in a nexus of confirmation relations—is to neglect the epistemic achievement of skilled coping that made that observation possible.

Skills of scientific observation are duly analogous to virtue: both are brought to maturity by habit—in some cases, by habit honed in very particular social contexts; both are accomplishments that, notwithstanding their contextual conditioning, can count as latching on to real features of the moral/empirical world. So let us suppose that Herschel's capacity to recognize comets, and similar capacities of other scientists, are epistemic achievements, best understood as the exercise of second-nature competences. If Traditionalism lacks the resources to understand them as such, Traditionalism is an incomplete epistemology of science.

Against such a charge of incompleteness, the Traditionalist might well protest that, while surely praiseworthy, Herschel should be viewed as a tank of evidence fueling a confirmation engine. Traditional epistemology aims to investigate the mechanics of the engine, not the composition of the fuel. Even if successful data gathering is an achievement promoted by the exercise of second-nature competence, the *epistemic* aspect of science has to do with what is done with the data and takes the form of obedience to perfectly comprehensive rules of scientific inference. So in the face of the charge that they neglect second-nature recognitional capacities, Traditionalists could well confess both that they do and that they never meant not to.

I'll meet the Traditionalist protest with a pair of quasi-historical examples, meant to suggest that epistemic good conduct requires more of scientists and their communities than diligently updating degrees of belief in response to new evidence. If the suggestion is true, *epistemically* significant features of scientific practice escape the notice of Traditional epistemology of science.

I'll call the first example, which is inspired by Alison Wylie's (1997) discussion of 1983 work by archaeologist Joan Gero, the Bison-Mammoth Construct. It is the 1970s. Edgeware analysis—using techniques including microscopy to study the physical morphology of archaeological relics with a view toward identifying their likely uses—is a feminized subdiscipline of archaeology. It is also low-status—low enough that the mainstream is slow to take up evidence from edgeware analysis. Instead, the mainstream is busy articulating the Man-the-Hunter paradigm, theorizing about bison and mammoth hunting practices and the social, technological, linguistic, and cognitive adaptations they required, as well as collecting evidence (for instance, from kill sites) informing these theories. A puzzle arises for the Man-the-Hunter paradigm: when the big game that hunters relied upon (according to the paradigm) for their subsistence went extinct, how did humans following suit? The (ignored) testimony of edgeware analysis: many of Man-the-Hunter's tools served foraging purposes; ergo Man-the-Hunter wasn't in fact subsisting solely on bisons and mammoths. The puzzle dissolves, and the paradigm that frames it is revealed to be incomplete.

The Bison-Mammoth Construct is the story of a community that persisted in embracing a theory longer than it should have. It persisted even after evidence challenging the theory surfaced, as it were, offstage from where the tastemakers were theorizing. *Even if* all members of the tastemaking community diligently updated their credences by conditionalizing via Bayes's rule on the evidence at hand, they were engaging in *epistemic bad behavior*. Enabling this epistemic bad behavior was the fact that the community was striated by status in such a way that its status hierarchies induced hierarchies of credibility that retarded or prevented relevant evidence from informing belief dynamics.

The foregoing diagnosis of epistemic bad behavior exhibited in the Bison-Mammoth Construct is a normative epistemological claim, and an important one. Yet it is a claim liable to elude the grasp of Traditionalism. The claim that status hierarchies have epistemic costs attributes *epistemic* significance to *contingent social structures* (ones that happen often to be indexed by race, gender, etc.). If status hierarchies carry epistemic costs, then we can't understand everything we ought to understand about empirical justification if we ignore social structures and the contingencies they condition.

The Bison-Mammoth Construct also illustrates a moral Peter Railton draws lucidly:

Comforting Bayesian thoughts about the influence of prior probabilities washing out in the long run require not only a very long run, but also a short run full of *many more alternative hypotheses* and *much more active conditionalization* than anything we actually see in scientific practice. (1994, 77; emphasis added)

Taking the moral to heart, we might suspect that Traditionalism is an incomplete epistemology of science. It's incomplete because it lacks the resources to understand as *epistemic* contributions efforts that *in the short run* enrich the collection of alternative hypotheses under examination, thereby facilitating more active conditionalization.

Even if Traditionalism is incomplete, do we have reason to expect that second natures modeled on virtue might help to deliver *epistemic* goods that Traditionalism can't? I think we can find support for this expectation in my second quasi-historical example, Swashbuckling Primatologists, which is inspired by the work of Donna Haraway (1991) and Sarah Hrdy (1986). Swashbuckling post–World War II primatologists are entertaining hypotheses about the crucial role that male-dominance hierarchies play in primate social structures, and evaluating those hypotheses against evidence collected from field studies, during which those primatologists paid most of their attention to manifestations of male aggression. They provisionally accept what, according to rules of rarefied warrant, are, in light of the data collected, the best of these hypotheses. A subsequent generation of primatologists, many female, many working in the wake of the women's movement, articulate alternative hypotheses about social structures, ones according significance to troop members who aren't adult males. These hypotheses are evaluated against observations of a broader range of troop dynamics than the swashbuckling primatologists made, and are, *according to the Traditionalist's evidential norms*, better supported by those observations than the swashbuckling hypotheses.

The epistemic moral of the Swashbuckling Primatologists example is that interests and habits, second natures, can play a positive role in the juggernaut of conjecture and refutation. The moral so stated concedes to the Traditionalist that valid scientific inference obeys rules of rarefied warrant. It nevertheless resurrects the epistemic significance in the context of justification of factors usually buried in the context of discovery and corrects for the exclusivity of the Traditional focus on rarefied warrant by suggesting that contingent histories can constructively interact with inferential engines.

In the Swashbuckling Primatologists example, what recommends the later over the earlier hypotheses is that they are better supported, not that they are unbiased (or less or differently biased). But not even religious adherence to norms of valid epistemic warrant will justify a community in choosing a hypothesis it hasn't thought of, supported by evidence it hasn't collected, over a recognized hypothesis supported by available evidence. The beliefs to which rational Bayesian engines drive convergence are only as well validated as our evidence and hypothesis sets are rich. Thus, the capacity to enrich those sets (vouchsafe evidence, imagine alternative hypotheses) contributes as significantly to the justification of hypotheses as does following rules of valid scientific inference. Insofar as exercising competences consigned to the context of discovery endows us with such capacities, those competences qualify as "epistemic resources" in much the same way as does the capacity to implement an inductive logic. Not specific to Bayesian subjectivism, the point can be adapted to apply to any Traditional epistemology that withholds epistemic significance from "context of discovery" factors that can promote the reliability of scientific inquiry.

If it is the aim of inquiry to adopt the best possible hypotheses for the best possible reasons, diligent adherence to evidential norms does not accomplish the aim on its own; it does so only in conjunction with the cognitive and technical innovations that bring hypotheses and evidence alike within the purview of those norms. Insofar as second-nature capacities for such innovations promote the aim of inquiry, they are as genuinely epistemic capacities as is the capacity to follow rules. And examples already given make it plausible that there are social/ scientific circumstances under which second-nature capacities can and do promote the aim of inquiry.

I'll close this section by briefly making explicit the connection between epistemic second natures and the possibility of a *feminist* epistemology of science (see Ruetsche 2004 for elaboration). I say "make explicit" because the suggestion is implicit in the quotation from Haraway about what Phyllis Jay *literally physically saw,* as well as in the Swashbuckling Primatologists example. To forge the connection, start with the view that second-nature capacities for cognitive and technical innovations promote the aims of inquiry and thereby deserve to be hailed as epistemic capacities. Append to that view the claim that there have been and might be domains of inquiry enveloped in such social situations that the gender and social situations of inquirers and potential inquirers—their second natures—affect (but in less dramatic versions of a feminist epistemology, fall short of determining) the capacity for cognitive and technical innovations. The result is a picture of how gender can matter to the epistemic dimension of inquiry, and matter in a way an epistemologist concerned only to articulate evidential norms might miss. The picture affords an understanding of how gender (and other contingencies) can matter substantially to inquiry. Because nothing about the picture cancels the broadly empiricist requirement that *evidence* is essential to empirical validation, the connection to gender is forged without reducing inquiry to wish fulfillment.

Engineering Science

The epistemology of science I'm calling Traditionalism aims to articulate the contours of rarefied warrant, an epistemic achievement attained by conforming to a calculus of rationality transparent to our rational first natures, and so an epistemic achievement to which differences between aspiring knowers make no difference. In the previous section, I've used the model of Aristotelian virtue, understood as a second-nature capacity, to propose that we recognize other sorts of epistemic achievement: epistemic achievements to which differences between knowers, their contingent histories, and their contexts are relevant.

The epistemology lying in the background of this proposal is broadly *externalist*. What qualifies second-nature capacities *as* epistemic is their efficacy in promoting aims of inquiry, such as sorting empirically adequate beliefs from empirically inadequate ones and enhancing the variety and the force of the evidence we have for the hypotheses we embrace. The epistemology lying in the background of the proposal is also *social* in at least two senses.

First, it recognizes epistemic achievements conditioned by social context. An example of an epistemic achievement conditioned by social context is the capacity to imagine hypotheses about primate social structure that aren't framed in terms of male-dominance hierarchies. The externalist grounds for counting the capacity as epistemic are that a community where the capacity is exercised collects more and better evidence than a community from which the capacity is absent. The community enhanced by the capacity entertains a wider variety of alternative hypotheses suggesting experimental approaches that generate a richer supply of evidence. A reason to count the capacity as conditioned by social context is that, as a matter of historical fact, scientists who had had their consciousnesses raised by the women's movement were able to imagine alternatives to male dominance as a central explanatory trope, alternatives that did not occur to scientists to whom the centrality of that trope seemed so natural as to be beyond question. It's plausible that the social locations of scientists made a difference to their capacities to contribute alternative hypotheses that challenged underlying assumptions of received models.

The proposal that second-nature epistemic capacities operate in the sciences is supported by a background epistemology that is social in a second sense. The background epistemology treats scientific communities as relevant epistemic units. For instance, the framing epistemology faults the Bison-Mammoth community for being so striated by status that it was slow to take up salient evidence recognized by outlier subcommunities. For another instance, the framing epistemology lauds as epistemic the second natures whose presence in a scientific community better enables that community to discharge its epistemic duties.

The social dimension of the epistemology of science framing the proposal developed here invites a question of epistemic design. *How would a scientific community be organized so that it might make highly effective use of epistemically relevant second natures?* This is a question of which Aristotle would approve. He considers "politics . . . pre-eminent among sciences, because it is responsible for organizing the state in such a way that its citizens become virtuous and that it it-self attains its Good" (II.i). A well-designed *polis* fosters virtue among its citizens, and the virtue of the citizens constituting a *polis* enables that *polis* to thrive. Our epistemic design question attributes to the epistemic engineer a preeminence analogous to that of Aristotle's politician. The epistemic engineer is responsible for organizing science in such a way that its practitioners bring to bear second natures that enable science to attain its Good, which we are supposing to consist in developing empirically cogent understandings of the natural and social world.

A possible tension lurks in the design task facing Aristotle's politician. Exercising virtue, a creature is supposed to attain the kind of thriving appropriate to it. But what guarantees Aristotle's politician that by designing a *polis* that delivers its *individual* citizens to thriving on their (human) terms, one thereby delivers the *community* consisting of those individuals to *its* overall Good? The balance of this section will discuss similar tensions lurking in the design task facing our epistemic engineer. A strategy for relieving them, suggested by the work of Helen Longino and congruent with the story told here, is to reconceptualize the Good of science in design terms.

The examples presented in the previous section indicate that a well-designed scientific community requires a wide variety of (at present often uninstantiated) broader social conditions to obtain—a point to which I'll return. The Swashbuckling Primatologist example suggests that in a well-designed scientific community, a relevant heterogeneity of second natures will participate.[5] And the Bison-Mammoth Construct example alerts us that the mere presence in a scientific community of a relevant heterogeneity of second natures may not be enough. The relevant heterogeneity of second natures must also be *enfranchised*, in the sense that their deliverances get sufficient uptake to affect community-wide processes of hypothesis development and evaluation. A community striated by prestige may frustrate such uptake.[6]

The design desiderata of *representation* and *enfranchisement* recall Helen Longino's epistemic ideal of a "cognitive democracy" (1990), a community characterized by equality of intellectual authority. And Longino's approach to the epistemology of science helps to illuminate connections, raised by the design question, between individual second natures and the Good of science. Longino urges us to understand epistemic accomplishment (knowledge, objectivity, and the like) in the sciences not as the accomplishment of individual unconditioned subjects but as an accomplishment of *communities* of *conditioned* subjects:

> Scientific knowledge is constructed not by individuals applying a method to
> the material to be known but by individuals in interaction with one another in
> ways that modify their observations, theories and hypotheses, and patterns of
> reasoning. (1992, 111)

Translated into the language of the present essay, Longino's proposal is to understand scientific knowledge as resulting from the right sorts of interplay of second
natures.

But what is the right sort? It might seem that, having abandoned a constricted
calculus of rationality as the sole arbiter of epistemic achievement, we have no
way to draw the sorts of distinctions that would enable us to answer this question. Longino denies this appearance. Holding that "community level criteria
can . . . be invoked to discriminate among the products of scientific communities,
even though context-independent standards of justification are not attainable"
(1992, 112), Longino proposes to compare, not hypotheses with respect to their
objectivity, but communities, with respect to their capacity to produce objective
beliefs. A distinctive account of scientific objectivity emerges from the comparison:

> The greater the number of different points of view included in a given commu-
> nity, the more likely it is that its scientific practice will be objective, that is, that
> it will result in descriptions and explanations of natural processes that are more
> reliable in the sense of less characterized by idiosyncratic subjective preferences
> of community members than would otherwise be the case. (1990, 80)

What Longino terms *transformative criticism* is an important means of exposing
idiosyncrasies and promoting objectivity as she understands it. Transformative
criticism is criticism that exposes and targets widely shared background
assumptions (1990, 76). It "requires the presence and expression of alternative
points of view" (1992, 111). And "a method of inquiry is objective to the degree that it permits transformative criticism" (1990, 76). Translating again into
this essay's vernacular, transformative criticism, and so objectivity as Longino
understands it, is sustained in a scientific community designed to represent and
enfranchise heterogeneous second natures.

Naming the Good of science "objectivity," Longino identifies it procedurally, in terms of a community structure permitting and promoting transformative criticism. Following Longino, we should be reassured that our epistemic
engineer's design task won't come to ruin on tensions between individual scientific thriving and community-wide scientific thriving. On Longino's account, a
scientific community organized to enable its participants to realize, represent,
and enfranchise their epistemic second natures is ipso facto a scientific community that realizes its own Good.

This is hardly to say that the approach to the epistemology of science here presented is without tension. A working assumption that does the critical work of qualifying second natures as *epistemic* is that whatever promotes uncontroversial epistemic values (such as empirical adequacy or convergence to the truth) is thereby an epistemic capacity, indeed an epistemic virtue. Considering the working assumption from the perspective of the design question has implications, some of which are surprising, for what virtues are epistemic.

What virtues are epistemic? On this telling, those second-nature capacities whose full realization in a well-designed scientific community promotes that community's pursuit of transformative criticism and hence of objectivity procedurally construed. The pursuit by a community of transformative criticism arguably requires of its members a suite of individual-level capacities—the humility to take the testimony of others into account, the patience to try to understand evidence arising from approaches imperfectly aligned with one's own, the trust to expend time and energy updating on evidence contributed by others. All of these individual-level capacities could well promote community-level pursuit of transformative criticism. And that would make humility, patience, and trust—prima facie moral virtues—epistemic virtues as well.

Peter Railton has suggested a less uplifting variation on the story just told. Like Longino, Railton reconceives objectivity in terms of appropriately designed scientific communities. He acknowledges that science as we know it is commodified, intent on prediction and control, intensely competitive, conducted by groups "led by dominant principal investigators—usually male—who are overcommitted to their research program" (1994, 79). Then he observes that *just such a design* is one you might adopt for a community "whose forms of interaction . . . might selectively reward reliability in belief formation" (81). You might adopt it because reliability fosters prediction and control, and hypercompetitive, overinvested groups competing for the monetary awards subsequent upon identifying effective strategies of prediction and control "might better fit the bill of producing novel forms of feedback that can frustrate subjective projection" than disinterested inquiry does (82). In the context of the design strategy Railton explores, the individual-level capacities that promote the collective pursuit of objectivity include arrogance and overconfidence. Moral vices emerge as epistemic virtues![7]

Noticing *that* epistemic virtues matter affords a way into—without settling!— the question of *which* virtues matter. The conflicting derivations just produced, one of arrogance and the other of humility as an epistemic virtue, suggest that background social conditions, assumed as given, could make a difference to how to answer what I've been calling the design question. But—like Aristotle's politician—we needn't take background social conditions to be given. We can also ask: what background social conditions must obtain in order to permit the organization of a scientific community that makes efficient use of epistemic second natures? Supposing we could determine an answer to this question, it's not obvious that we would like it. A social organization that fully promotes epistemic values could well sacrifice others' values, such as fairness or equity (think of Plato's *Republic*). And even if the social conditions that, say, fully promoted transformative criticism *also* measured up as just in other respects, there could well remain moral and political qualms about the permissibility of steps that might bring those social conditions about.

I'm rooting for social justice, scientific objectivity, and individual thriving (both moral and epistemic) to turn out to be mutually reinforcing goods, and moreover goods that are both practically and ethically attainable. I don't know that they are. I have tried to suggest that one way to think about whether they might be is to expand our notion of scientific reason to incorporate epistemic capacities modeled on Aristotelian virtue.

Notes

* For support, inspiration, and feedback, I am grateful to the editors, to Karen Frost-Arnold and the other participants in the workshop on which this volume is based, and (despite their Traditionalism) to two of my favorite teachers: John Earman and Wes Salmon.
1. Joyce 2011 and Weisberg 2011 are two recent surveys of variations on the thought.
2. Not for the first time. The present essay is a short-form remix of Ruetsche 2004, which uses epistemic second natures to characterize and evaluate a variety of theses in feminist epistemology.
3. I say "almost" because agents who are dogmatic—that is, agents who assign degrees of belief of 0 or 1—can derail "washing" out results.
4. All Aristotle quotations are from the 1926 Horace Rackham translation.
5. For some recent empirical work on what sorts of heterogeneity are relevant to what sorts of inquiries, see Page 2019.
6. For formal results supporting this contention, see Hegselmann and Krause 2002, which Ruetsche 2019 discusses in this connection.
7. I'm not sure that Aristotle would count arrogance as a vice—see his discussion of the moral virtue *megalopsychia* in the *Nicomachean Ethics*, IV.3–4. But I would.

References

Aristotle. 1926. *Nicomachean Ethics*. Translated by Horace Rackham. New York: G. P. Putnam's Sons.

Bacon, Francis. 1960. *The New Organon and Related Writings*. Edited by F. H. Anderson. Indianapolis: Bobbs-Merrill.

Cartwright, Nancy. 1997. *Nature's Capacities and Their Measurement*. Oxford: Clarendon Press.

Earman, John. 1992. *Bayes or Bust? A Critical Examination of Bayesian Confirmation Theory*. Cambridge, MA: MIT Press.

Galison, Peter. 1997. *Image and Logic: A Material Culture of Microphysics*. Chicago: University of Chicago Press.

Hacking, Ian. 1983. *Representing and Intervening: Introductory Topics in the Philosophy of Natural Science*. Cambridge: Cambridge University Press.

Hanson, Norwood Russell. 1960. "Is there a logic of scientific discovery?" *Australasian Journal of Philosophy* 38, no. 2: 91–106.

Haraway, Donna. 1991. *Simians, Cyborgs, and Women: The Reinvention of Nature*. New York: Routledge.

Harding, Sandra. 1986. *The Science Question in Feminism*. Ithaca, NY: Cornell University Press.

Harding, Sandra. 1991. *Whose Science? Whose Knowledge?* Ithaca, NY: Cornell University Press.

Hegselmann, Rainier, and Ulrich Krause. 2006. "Truth and Cognitive Division of Labor: First Steps Towards a Computer Aided Social Epistemology." *Journal of Artificial Societies and Social Simulation* 9, no. 3: 10.

Hrdy, Sarah. 1986. "Empathy, Polyandry and the Myth of the Coy Female." In *Feminist Approaches to Science*, edited by Ruth Bleier, 119–146. New York: Pergamon.

Joyce, James. 2011. "The Development of Subjective Bayesianism." In *Handbook of the History of Logic*, Vol. 10, *Inductive Logic*, edited by Dov Gabbay, Stephan Hartmann, and John Woods, 415–76. Amsterdam: Elsevier.

Longino, Helen. 1990. *Science as Social Knowledge*. Princeton: Princeton University Press.

Longino, Helen. 1992. "Subjects, Power, Knowledge: Prescriptivism and Descriptivism in Feminist Philosophy of Science." In *Feminist Epistemologies*, edited by Linda Alcoff and Elizabeth Potter, 101–120. New York: Routledge.

McDowell, John. 1976. "Virtue and Reason." *Monist* 62, no. 2: 331–50.

Page, Scott E. 2019. *The Diversity Bonus: How Great Teams Pay Off in the Knowledge Economy*. Princeton: Princeton University Press.

Railton, Peter. 1994. "Truth, Reason, and the Regulation of Belief." *Philosophical Issues* 5, no. 1: 71–93.

Ruetsche, Laura. 2004. "Virtue and Contingent History: Possibilities for Feminist Epistemology." *Hypatia* 19, no. 1: 73–101.

Ruetsche, Laura. 2019. "What Is It Like to Be a Woman in Philosophy of Physics?" In *Routledge Handbook of Feminist Philosophy of Science*, edited by Sharon Crasnow and Kristen Intemann, 397–408. New York: Routledge.

Salmon, Wesley. 1966. *The Foundations of Scientific Inference*. Pittsburgh: University of Pittsburgh Press.

Weisberg, Jonathan. 2011. "Varieties of Bayesianism." In *Handbook of the History of Logic*, Vol. 10, *Inductive Logic*, edited by Dov Gabbay, Stephan Hartmann, and John Woods, 477–552. Amsterdam: Elsevier.

Wilson, Mark. 2018. *Physics Avoidance: And Other Essays in Conceptual Strategy*. Oxford: Oxford University Press.

Wylie, Alison. 1997. "The Engendering of Archaeology: Refiguring Feminist Science Studies." *Osiris* 12, no. 1: 80–99.

10

Is "Failing Well" a Sign of
Scientific Virtue? *

Jutta Schickore

Introduction

Several science educators, science writers, and scientists writing for general audiences have claimed that failure should be more appreciated in science. Failure is unavoidable in everyday scientific practice and is not an indication of "broken science" (see, e.g., Parkes 2019; Zaringhalam 2016). By emphasizing that failing is frequent and to be expected, these commentators seek to ensure that nonscientists form a more adequate view of how science really works, and that science students come to have realistic expectations about their future research activities.

The neurobiologist Stuart Firestein has recently promoted an even more positive attitude toward failure in science in his popular book *Failure: Why Science is So Successful*. Failure, Firestein tells us, is "both an inevitable and a *desirable* part of science" (Firestein 2015, 19; emphasis added). We tend to think of science as astoundingly successful because it is so good at overcoming failures. Firestein challenges this view, seeking to highlight that science is composed "mostly of ignorance and failure—with perhaps a dash of accident or serendipity thrown in" (3). On this view, science is so successful because it is so *good* at failing. Firestein claims that this is "not a secret at all: it's general knowledge, inside of science" (4).

I am sympathetic to the general agenda of promoting, among science students and the general public, a more realistic understanding of how science really works, but I doubt that shifting the emphasis from valuing success to valuing failure serves the purposes it is intended to serve. In this chapter, I discuss several questions: Is failure an *inevitable* part of science? Is failure a *desirable* part of science? Do *scientists* have an adequate understanding of how science works—and the role of failure in it—and does it matter if they do? I begin by distinguishing different kinds of failures: the incapability to adhere to a rule, the incapability to devise a (new) approach, and the dead end. Second, I suggest that analysts of science who seek to provide accounts of how everyday science works need to move beyond the dualism of failure and success. Much of everyday scientific practice is

Jutta Schickore, *Is "Failing Well" a Sign of Scientific Virtue?* In: *Science, Technology, and Virtues.* Edited by: Emanuele Ratti and Thomas A. Stapleford, Oxford University Press. © Oxford University Press 2021. DOI: 10.1093/oso/9780190081713.003.0011 not well described as either achieving success or failing to do so. Very often, it is unclear whether research outcomes are one or the other.

Third, I examine the peculiar notion of "desirable" failure or "failing well." We will see that Firestein's notion of "failing well" is a repackaging of the old idea of the maverick genius. One's peers, acting as referees, reviewers, and administrators, might reward or seek to restrain innovative thinking. Implicit in Firestein's praise of failure is the allegation that scientists and science administrators typically do the latter while they should really do the former.

Fourth, I contend that "inside of science" there are also misunderstandings of how science really works, and these misunderstandings have an impact on the functioning of science. Revaluing failure will not help to remove these problems. To support this claim, I examine working scientists' understanding of good and bad scientific practice, and of failing and succeeding. I draw on a number of interviews and analyze the expressed attitudes toward failure and toward everyday challenges in science more generally. Contrary to Firestein's suggestion that the ubiquity of failing is "general knowledge, inside of science," I argue that in interviews with working scientists, failure typically has a negative connotation; it is typically not seen as a desirable part of science. Scientists themselves usually place the emphasis on success in just the ways that science educators and science commentators such as Firestein want to combat. Reacting to the social and institutional constraints on scientific practice, scientists stigmatize failure as personal defeat, and they see everyday scientific practice as a practice where failures ought not to occur.

In the concluding section, I discuss whether it matters—and if so, why it matters—whether scientists are right or wrong about how science works. I argue that it does matter because the stigmatization of failure and the underappreciation of the openness of everyday science have problematic implications not only for the public image of science but also for scientific practice itself and for the communication of research outcomes *within* science. Not only do scientists need to be aware of the social dimensions of failure and success, but they also need to recognize that everyday practice is more open and flexible than traditional philosophical or popular notions of everyday research suggest. Scientists who appreciate the social dimensions of failure and success and openly acknowledge the flexibility of scientific research practice are likely to be able to deal more prudently with the institutional pressures endemic in science.

A Note on the Interview Data

My account in this chapter is based in part on interviews with working scientists. These interviews were conducted in 2011 at Indiana University, Bloomington, and at the Indiana University School of Medicine, Indianapolis.[1] The database consists of thirty-eight interviews. At the time, all of the interviewees were involved in empirical, mostly experimental research. The interviews were semi-structured yet thematically focused, as researchers were asked directly about errors, failures, negative results, and their communication (or lack thereof).[2] The interviews were reexamined for the purposes of this chapter.

Sixteen respondents were from the IU School of Medicine (fifteen as faculty members and one as a postdoctoral researcher). The remaining twenty-four respondents were from IU Bloomington and represented sixteen departments and schools across the academic spectrum. Four respondents were postdoctoral fellows at the time, fifteen were assistant professors or scientists, five were at the associate rank, and sixteen were at the full or senior rank. In this chapter, I use quotes from twenty interviews. The fact that the interview population is quite small precludes interpreting the results as statistically significant. Nevertheless, the interviews are enlightening because the in-depth accounts provide a wealth of detailed information, and they have merit because a diversity of departments is represented and a multiplicity of interpretations of certain phenomena is evident.[3]

Another limitation is that the empirical research initiated a learning process for myself as analyst. The initial interview guide included a preconceived list of terms.[4] The interviewees were asked to comment on these terms, and they were encouraged to add their own reflections and raise additional issues. From the analysis of these responses and reflections, new, more specific, and targeted questions have emerged that could be asked in another round of interviews with a new group of scientists, which will lead to even more intricate questions.

The chapter cannot and does not aspire to be a full-blown empirical study of scientists' attitudes and behaviors. However, analyzed alongside other materials—such as Firestein's book and philosophical, sociological, and psychological studies of failure and uncertainty in research practice—the interviews are indicative of the manifold issues related to uncertainty, succeeding, and failing, or failing well, in science, and they challenge our philosophical understanding of scientific practice in various ways.

Two Kinds of Instructive Failures

What, if anything, is desirable about failure and failing in science? Is there more to it than just the cliché that we should not expect to get it right at the first trial, that we should persevere and try again—and again, and again? If failure is not just inevitable but is indeed desirable, is it correct that "inside of science" everyone is aware of this? Is the good scientist the one who consciously embraces failing or who knows how to "fail well"? The questions are difficult to answer. To be able to designate something as "failure," we need to have an understanding of what "success" might look like, but the concept of failure is elusive—as is the concept of success.

The *Oxford English Dictionary* offers three definitions of failure, all of them negative. There is failure as omission: failing to occur, to be performed, or to be produced; omitting the performance of something due or required. There is failure as breakdown: becoming exhausted; giving way under trial; breaking down in health; declining in strength or activity; bankruptcy, et cetera. There is the failure of not meeting one's goal: failing to effect one's purpose. A thing or person that proves unsuccessful is also a failure in this sense.

The connotations of the verb *fail* listed in the *OED* are also all negative. Failing could mean to be or become deficient, exhausted, extinct; to flag, to prove deficient; not to render the expected service; to lack, to fall short, to disappoint, to leave undone, to miss the mark, to err, to be unsuccessful in an attempt, or to become bankrupt.

It is easy to think of failings in science, small and large, that fit the scope of the dictionary definitions. Scientists may omit to clean a test tube, and as a result, the chemical analysis fails. They may not be able to bring a project to a successful conclusion because the funding is running out. A researcher may decline in activity due to administrative duties, and so on. Again, what comes to mind are negative instances: things and actions that are an inevitable ingredient of everyday practice, yet are undesirable or should be avoided.

What, if anything, could be positive or desirable about failure? There are countless slogans, maxims, and proverbial sayings in praise of failing, such as "The only real mistake is the one from which we learn nothing" and "A person who never made a mistake never tried anything."[5] Newspaper articles and magazines praise the value of failing as a life experience (e.g., Babur 2018; Vallery 2016). They suggest an obvious point—namely, that failures can play a productive role in education. In educational contexts, error and failure can serve as a grooming tool. Good educators are able to turn failures into teaching moments, and they may even intentionally create the conditions to make students fail (Athanassoulis 2017). These kinds of instructive failures—regardless of whether they are instructor-designed—may help you learn to do things right and teach you to deal with the inevitability of failure. You fail in your attempt to follow a certain rule or to apply a principle *correctly*; you are instructed to try again; you get better at it. Instructive failures in this sense are means to an end; they "help on the road to virtue" (Athanassoulis 2017, 347).

These ideas can be easily applied to science education as well. Starting in grade school, students should be encouraged to practice doing research in order to hone their skills and to get better, rather than to try once and give up or, worse, to try to learn science by memorizing facts.[6]

There is a complex set of motivations, assumptions, and expectations tied to these efforts to normalize the experience of failing. The underlying worry is that students who experience failure in science classes might get too easily discouraged and might not even consider taking up a science career. Ultimately, these efforts are intended to ensure the continuity of the current institution of science, both in terms of effective recruitment into science and in terms of preserving the existing structures and practices. Arguably, in these instructional contexts, failure is a positive thing, at least in the eyes of the educators, and, so an educator may hope, in the eyes of every prudent student as well. Having failed and tried again in an educational environment will, one may further hope, result in more skillful, more resilient individuals who can do everyday science competently and confidently, with tenacity and perseverance.

Incidents of failing could also be seen as instructive in educational contexts if we consider how difficult it is for researchers to develop truly novel models or innovative theories. Scientists who were ultimately very innovative have had to overcome setbacks, periods of stagnation, and want of success; even famous scientists have had to work hard to overcome adversity and to develop a new approach or theoretical framework (Nickelsen and Graßhoff 2008; Steinle 2016).

The searching, erring, and failing that preceded theoretical breakthroughs are often shared in autobiographies, memoirs, and personal reflections. François Jacob famously used the term "night science" for scientists' struggles to generate new knowledge: "Night science wanders blind. It hesitates, stumbles, recoils, sweats, wakes with a start" (Jacob 1998, 126). Science educators may use stories of struggle to make science students aware of the toils and labor involved in situations of theory choice. This will give the students a much better idea of the challenges ahead than will teaching them just the finished products of research. These stories of struggle also convey a positive side to failing: students will learn that effort, drive, and perseverance are important psychological conditions of scientific achievement, at least as important as ability and talent—and that eventually their failures may turn into successes.

Notably, however, these two notions of instructive failure are fundamentally different. The first kind of instructive failure is a failure to follow a known and generally acknowledged rule or to miss a specific goal or target. The student who is learning to do research may fail to follow an experimental protocol or fail to do a Fourier transform, instances in which one arrives at a clear sense of what it means to succeed. The instructor will know what to do and what is right and should be able to teach the student how to do it right next time.

The stories of struggle, setbacks, and stagnation, by contrast, are about scientists finding themselves in situations where nobody knows what to do or what is right. In times of struggle, the point is that it is unclear *how* one might turn the failure into a success, what "success" might even mean, or even *whether* a solution exists in the first place. As Jacob put it: "*There is no way to predict whether night science will ever become day science; whether the prisoner will emerge from the darkness*" (1998, 126; emphasis added).

The philosopher Giora Hon has advanced a conceptual distinction that captures the two types of instructive failures. Hon distinguishes between a "mistake" as a violation of a rule and an "error" in the sense of going amiss without knowing where to go or what is right (Hon 1995). The first of the two kinds of instructive failures just discussed is like Hon's "mistake": it is a failure to perform according to established rules. This failure can be corrected by pointing to these rules. An example of a failure or mistake in this sense would be a typo. Error, in Hon's understanding, is the process of erring, of wandering around, of having lost one's way, of stumbling blindly.

All these distinctions—mistake and error, the two types of instructive failure, night science and day science—map onto older accounts of scientific change as they were advanced by Kuhn, Lakatos, Laudan, and others. According to these philosophers, scientific fields go through periods of more or less radical change whereby new paradigms, research programs, or theoretical frameworks are developed and established. These periods are marked by the second type of instructive failure: the "night science"; the erring, stumbling, and wandering; the not knowing where to go next. These periods of theory change are interspersed with extended periods of work within an accepted paradigm or theoretical framework. In these periods, mistakes can be made—the failures to perform according to established rules are the equivalent of not conforming to the demands of a research program or of not being able to solve a puzzle within a paradigm.

Erring or "night science" is the equivalent of the inability to come up with a new research program or paradigm. In these situations, an instructor's role can only be psychological. The instructor can offer counseling and support but no rules for discovery. Intricate philosophical discussion about the possibility of a logic or algorithm of discovery have shown that there are no recipes for the generation of new theoretical frameworks: "Planned searching can be of little use if one has not yet found the idea that brings deliverance, which must be guessed before it reveals itself as such to the surprised enquirer" (Mach 1976, 214). Even though some strategies for systematic searching are characteristic of these periods, including parameter variation and the search for "simple cases" (Steinle 1997), applying these strategies does not guarantee that new knowledge will be generated. In hindsight, we understand that in many experimental contexts we now regard as generative of new knowledge, such strategies were involved. Yet strategies of systematic searching or "exploratory" experimentation are not a recipe for cooking up new ideas, nor do they guarantee that the new ideas thus generated will withstand further scrutiny. Instructors may refer to such situations not to teach methods of discovery, but only to provide general encouragement. As such, they have an important role to play, however. Empirical studies suggest that stories of the struggles of famous scientists ("Even Einstein struggled . . .") do have a positive impact on students' efforts and, ultimately, on their achievements (Lin 2016).[7]

Hindsight and Iconoclasm

Instructors, science students, and indeed struggling researchers may take comfort in the fact that, in hindsight, many projects, although considered failures at the time, were constitutive for later scientific advancements. Without the earlier projects, later advances might not have occurred.[8]

This is how Harry Collins now interprets Joseph Weber's work on gravitational waves (Collins 2017). Without Weber's failure in the 1970s to detect gravitational waves, the recent successes would not have happened. So, *was* Weber's project really a failure? The answer depends on one's historiographical perspective. There are good reasons to say yes: after all, he did not bring his experiments to a successful conclusion, and he could not convince the scientific community at the time. But there are also good reasons to say no: after all, later successful detections of gravitational waves built on his work. We may call the former historiographical perspective on failure situational because in our evaluations we are considering the verdict of the peers at the time. The latter historiographical perspective on failure could be called consequence-oriented because in our evaluations we are considering the long-term consequences of a research project to determine whether that project, or some part of it, was a failure or a success.

The distinction between situational and consequence-oriented perspectives is tenuous, however, because it crucially depends on the analyst's framework for reconstruction and timeframe. The question "When did a project (really) fail?" is as elusive as Peter Galison's questions about the successful conclusion of an experiment (1987): How (and when) do experiments end? Who decides? And how many failed projects are truly "dead ends" in the sense that *nothing* follows from them? Can we not, in most cases, point to *something* good or useful that came out of a project? Perhaps an attempted solution that failed in the past will be picked up at a later time. Different means—newly available technology, for instance, or a new interpretive framework for old, useless data—could make a difference and turn earlier failures into successes. Again, to answer these questions one would need to specify the meaning of "success," but it is as difficult to pinpoint what "success" means in science as it is to identify "failure."

Notably, this relational notion is still not the concept of "good failure" that Firestein has in mind. Firestein's notion of good failure takes us out of the historiographical and teaching domains and into actual research and science-policy contexts. In fact, it is exactly the opposite of the instructive failure that helps one to learn the ropes and to be tough and virtuous at the same time. It is also different from the failure in situations where one is stumbling and wandering around because no rules, exemplars, or frameworks are available for orientation and guidance. Firestein's good failure is iconoclasm: the *intentional* breaking or exploring of established rules, be they explicit or tacit. Iconoclasm in this sense is a feature of innovative theater, music, or painting. As the performance studies scholar Sara Jane Bailes has noted, experimental theater questions the rules of theater-as-representation, thereby exploring and perhaps surmounting its limitations.[9] Firestein (a scientist with a past in theater) invites us to see this notion of failure—exposing the limiting and constraining functions of rules—also as a hallmark of original, innovative science.

Firestein's (and Bailes's) notion of failure as iconoclasm is an attempt to promote creative criticism by pointing out how important the questioning of rules is for the advancement of science (and literature and theatrical performance). We are making science "better"—we are making *progress* in science—by failing to do business as usual, by going beyond it. This kind of failing is different from both the instructive failure, where the student fails to adhere to the established rules of everyday science, and the struggles of an Einstein, where there is nothing to guide one beyond the established rules (nothing but the rules of thumb of exploratory experimentation). Iconoclasm occurs when rules *are* in place and we deliberately explore or defy them. Again, we have here an old idea with a new label: he who is failing well is the maverick, the ingenious bad boy who advances science by being a nonconformist. (Feyerabend comes to mind.)

There is both something right and something problematic about this idea of desirable failure. Epistemically speaking, the oxymoron of "good failure" surprises and challenges our understanding of what it means to fail, which is second-level iconoclasm, so to speak. But the notion also obscures, because it suggests that we can learn something new from failure beyond what *doesn't* work; that we can expand our knowledge through failing. However, defying existing rules is one thing; coming up with alternative concepts and rules is another. Tearing down may make room for something new, but it does not, in and of itself, generate something new. Learning something positive *from* failure— beyond learning what does not work—can happen if we know all the possible options and we also know that one of them is correct. But science often is not like trying one of a bunch of keys and finding it doesn't work, knowing that the right key is in the bunch. Rather, one remains in the dark as to what to do next. Such failures may be suggestive, but one cannot turn one's failures to successes just by inspecting the failures. There is no secure path to new knowledge, no recipe for discovery.

The notion of "good failure" as celebration of iconoclasm obscures because it ignores the social-institutional dimension of establishing new rules. At stake is a plea to science administrators and funders not to restrain creativity unduly but rather to tolerate more original, creative research that goes against the grain of mainstream science and to provide the space, money, and time necessary to pursue such research. A number of scientists, often later in their careers, have called more directly for greater freedom of research in an attempt to recast the conditions for a productive scientific life. Firestein's idea of the value of iconoclasm can be recast as a critique of the conservatism of the scientific community or as a plea for more freedom in science, more freedom to tinker and to pursue sidelines with less competition for funding; all this on the assumption that more freedom would allow for serendipity and for "happy accidents" to occur (Meyers 2007).[10] A productive scientific life in this conception is one of freedom of choice of research avenues and freedom to pursue strange avenues; such a life is thought to enable scientific innovation and progress.

However, there is a social side to iconoclasm in the sense that the relevant community decides whether an iconoclast is a rule-breaker (and thus wrong) or a game-changer (and thus spectacularly right).[11] Freedom from the constraints of science administrators does not automatically mean freedom from the demands of peer evaluation. Perhaps, if limitless funding were available, peers would be more generous. Otherwise, the commitment to iconoclasm begets no real elevation of failure, but rather a mere shift of validation. The cynical rendering of this point would be that once it is considered desirable to be iconoclastic, those who do business as usual are failures. "Failure" is (and remains) a label for everything that others find undesirable; one is only trying to change the content of what should perceived as objectionable. Rule-abiding is now perceived as failure—or at least as not desirable.

Personal Defeat

I will now examine what scientists themselves say about failing and "blind wandering" in their research. When researchers in interview settings are asked directly what the term "failure" means to them, they do not say many positive things about it. In interviews, several individuals noted that "people don't like to fail" (researcher, medical education and pediatrics) and that they don't want to admit that they failed—"I mean, you don't want to say it" (associate professor, anatomy and cell biology). One said, "Human nature, being what it is, people don't stand up and go, 'Yep, I was wrong. You're right'" (associate professor, surgery). The overwhelming majority of comments regard failing as a personal defeat: the personal experience of falling short of one's goal, of being unable to perform a certain task, of being unable to get something to work as expected or at all.

Only one interviewee—notably, a researcher turned curriculum planner—echoed the science educators' notion of instructive failure that I have discussed. This interviewee commented on what she called the "self-esteem movement to let's make children feel good about themselves, regardless of what they've done." She suggested that such an approach to science teaching would lead to underperformance and called for just the kind of training that utilizes instructive failures. She observed that "if we are in a society that does not value failure . . . then we can't possibly think about learning from failure, and if we so need to succeed that we will change our standards, we will minimize our standards in order to succeed." She noted that in her capacity as an administrator she had "often thought about . . . how we could create a curriculum of failure and resilience; how we could force students to go out and fail and then have to respond to that failure and rise from that failure" (professor, instructional systems technology).

Other scientists, however, associated failing with potential threats to one's career and future employment. In science—our current science, with the current funding policies—only immediate successes are rewarded. When asked about errors and failures, one researcher said, "The first thing that comes to mind . . . is . . . failure to get tenure." He also mentioned "rejection of papers" (associate professor, telecommunication). Another gave as an example of failure "if a proposal or a grant funding application were to not be accepted" (research associate, physics). Rewards are not given for contributions to scientific projects on the grounds that they may later turn out to be a major scientific advancement.

Many responses conveyed the sense that the failing researcher is somehow flawed. One physicist said, "I've never seen a paper where the author said, you know, 'We have to revisit and start out anew because we didn't figure it out.' I think usually they wait until they publish when they have got the answer figured out" (researcher, medical physics and radiation oncology). Another scientist said that failing to find something has "more of a negative connotation, because it's not just that you didn't find something. It's sort of that it was your fault you didn't find something" (research scientist, education). A medical researcher reflected: "Sometimes things just don't work. And that's good to know . . . And I think in medicine people don't have that idea. Or I think people in science don't always have that idea of just because something doesn't work doesn't mean it's, you know, it's the whole Thomas Edison thing. 'I haven't failed 266 times, I've found 266 ways to not make a lightbulb.' And most people don't think that way" (researcher, medical education and pediatrics).

One physicist pointed out that people "have made a proposal to measure something . . . , and they set to work on it, and then it fails for some reason, *not necessarily under their control*. Very often the laboratories that depend upon accelerators to produce particles for experiments don't work as efficiently as they need to" (senior researcher, physics; emphasis added). Other interviewees were more ambivalent about whether *things* do not work or whether *experimenters* fail. One illustrated the ambivalence: "Failure, to me, would indicate something flawed or something in your research that affected the results, and this could be just naive assumptions that you made, and then [it] turned out that your results, that these assumptions, were wrong, so you failed to show whatever you were going for. It could also mean that you just completely chose the wrong methodology" (research scientist, informatics). For many interviewees, the main issue is not that *things* don't work (or don't work well enough) but that *they* failed to *get them* to work. One researcher said: "Failure has more of an implication that it was somehow a flaw in the design" (research scientist, education). These quotes about faults and flaws suggest that failure takes on a moral dimension.

Stigma

In the early years of laboratory ethnography in science and technology studies, sociologists Nigel Gilbert and Michael Mulkay examined scientists' comments about errors in their research. They pointed out the asymmetry of error ascription in the interviews they conducted. Their biochemists treated "correct belief" as the normal state of affairs and as unproblematic (in the sense of not requiring an explanation). Error, by contrast, was a failure to understand the real scientific issues (Mulkay and Gilbert 1982).

For these biochemists, error and failure were associated with ignorance and cognitive limitations: the others got it wrong, they had failed to understand. Typically, the interviewees ascribed these failures to others. Of the people that were interviewed for the present study, one—a particle physicist—expressed exactly this attitude, saying, " 'Error' as in, you know, someone did something wrong, you know, they really just didn't understand this technique" (research professor, physics). Another researcher working in science education said that "failure" is "used in terms of when an expectation isn't met, you know, either by the authors of that study or by somebody maybe in a review . . . trying to buttress their own argument by the lack of results of someone who's presenting a point of view that challenges theirs. So, you know, the failure of so-and-so . . . to find contradictory results just strengthens my case" (professor, curriculum instruction). The passages quoted in the previous section, however, suggest that thirty years after Mulkay and Gilbert, the cognitive notion of failure has become morally charged. There is a stigma associated with error and failure. Failure is a *bad thing*. If you fail, you must have done something objectionable; at the very least, you failed to do your duty as a scientist.

The stigma associated with error and failure appears to be so strong that scientists even shy away from accusing other people of failure, at least in public. One researcher said failure "is quite strong. . . . It's close to a subjective or it's close to a value judgment. I would think that it would be wiser to say 'unintended consequences' or 'the policy doesn't work as intended'" (professor, ethnography and education). In a similar vein, it was pointed out that in publications it is unseemly to use "failure" regarding one's own work or other people's works. A science policy researcher said, "The times that I would see that word are sort of more in editorial discussion pieces, either at the front end or in the concluding discussion sections of an article where somebody might say something like, 'The administration's failure to consider blah in the implementation of this policy may have led to the following results,' but not . . . when discussing results per se" (assistant professor, public and environmental affairs). The stigma associated with failure and error means it would be a bad idea to admit to them and in bad taste to ascribe failure to others. Instead, authors would be "dancing around it" (research scientist, education).

Negative Results and Failures

While the stigma surrounding projects that did not work can help explain why researchers are unwilling to share information about failed projects, one would expect that negative results are in a different category. Surely, one cannot reasonably expect that one's predictions and expectations will always be substantiated in experimental research, or in clinical trials for that matter. Is it not a relevant contribution to our knowledge that one treatment does not work better than another, or that a certain expected effect does not occur?

Indeed, one researcher stressed that "negative results . . . are good to know because then people aren't going to waste their time trying to do this" (researcher, medical education and pediatrics). A number of people said that repositories for negative results or outlets for reports on things that don't work would be extremely helpful—although very few had heard of existing negative results journals (see also Tsou et al. 2014). Notably, one medical researcher said that "editors don't like to see a lot of these negative results published because their perception is we want to have an impact factor, and our impact factor dictates who's going to publish in our journal. Impact factors can be influenced . . . by positive results, but they can be influenced by negative results as well, because if you cite a negative result paper it's going to get a lot of citations just as much as a positive one, and a lot of them don't see that." The researcher described this as a "caveat of the field" (associate professor, pediatric research and biochemistry and molecular biology). Another said negative research outcomes would be recast into positive results: "It's always, you know, 'We believe that this is the case and here's our great evidence.'... There will be an entire paper constructed around an argument against something else but the internal dialogue of the paper is, 'Here's what we think is going on and look, our evidence supports what we think is going on. And all of this does reject that over there.' What you don't see is 'Here's an idea and now we show clearly that that idea is wrong'" (associate professor, telecommunication). The medical researcher quoted earlier pointed out that negative results should not be regarded as an endpoint to one's project: "A lot of people will say that they created an animal and they manipulated a specific gene in that animal and their hypothesis was that 'When I manipulate this gene, I'm going to see the animal get cancer.' So they manipulate the gene, they wait for two years, no cancer develops. So they say, 'Well, that's a negative result. OK.' They'll publish that and dump it, but the problem is, even though that was a negative result, a lot of these guys never go and look for something else that could've occurred. So it's like they're not trying hard enough to find the answer. They just tried hard enough to find to validate their answer and not look at a broader sense, and try to think, 'Maybe there's more to this,' and you see that a lot" (associate professor, pediatric research and biochemistry and molecular biology).

It appears that negative results are stigmatized just as much as projects that don't work. One medical researcher described a clear case where (her own) negative results were devalued. She had tested a new, advanced piece of medical technology, impregnated catheters, and found no statistically significant difference in complications compared against standard catheters. She stressed that it was "extraordinarily difficult" to get the study outcomes published in an academic journal. Two journals said, "Well-designed study, well-executed study, but unfortunately failed to show any difference between the two catheters, and so good reviews on even how the paper was written, but not a priority for this journal at this time because it didn't show a difference between the two catheters."

The researcher noted that she found her finding very important, even though it was a negative result: "Why would you make a switch to that sort of catheter when there's data that says it doesn't make any difference? So I think while it is a negative result study, it is still so relevant because it can impact people's practice, and then maybe somebody else won't repeat the same study, with so few funds around for people to study now. So that's kind of what surprised me more than, well, they had wished that it had been positive result. ... And I guess I wonder how many studies there are that actually have negative results, and are they being given their due for those reasons that the negative is just as important to know ... as the positive" (professor, pediatrics).

Recent studies of publication practices in the medical sciences have drawn attention to the mechanisms by which pharmaceutical companies control and manipulate the publication of trial outcomes: gag clauses, suppression of negative results, ghosting, grafting, and so on (Krimsky 2003; Sismondo 2009). In the example just mentioned, the difficulties arose with peer reviewers for academic journals. Scientists themselves perceive this practice of obstructing the publication of negative results and failed projects as a problem. Nevertheless, once the researchers act in an institutional role as reviewers, and thus as gatekeepers, they change their perspective on communicating negative results and failed projects, as the case of the catheters indicates (see also Tsou et al. 2014).

A professor in education described another case of failing to publish a negative results study. In this case, it was a study on classroom technologies. As in the previous example, the study was "rigorous" and "could have been replicated." Nevertheless, it was rejected by three peer-reviewed journals. The researcher remarked, "None of those three journals were interested in publishing that article because it was a negative results article" (professor, instructional systems technology). The fact that she called it "a failure study" illustrates conceptual slippage between negative results and failure. Another interviewee explicitly noted this conceptual slippage: "People have been sort of conditioned that negative results is a failure of something" (researcher, medical education and pediatrics). If this is the case, then it is not surprising that negative results studies have the same fate as failed projects. The stories the interviewees share confirm this ambiguity of the word "negative" in the phrase "negative result."

In the context of comments about the failure to replicate something, "failure" takes on a moral connotation. When scientists report that they could not repeat another group's work, they insinuate not only that the other group did something *wrong* but indeed that they did something morally *bad* (see also Hangel and Schickore 2017, sec. 4.). It is notable as well that researchers in established fields perceive scrutiny of research practices through replication and other tools of critique as attempts to detect sloppy practice or, worse, unethical behavior. One interviewee mused, "Sometimes I think, well, wouldn't it be nice to have a theory about the way the world works, then test it multiple times, or you replicate it. . . . but there's really not much incentive to do that. Sometimes I've tried to do it and . . . I've learned from my senior colleagues that [you] don't do that because it . . . can get you into trouble" (associate professor, political science).

Young Fields and Open Discussions of Methods

Scholars in young research fields are much more tolerant than others of failures and things that do not work, while advocating for changes in communication practices.[12] For instance, a digital humanities scholar recalled an essay in that field, "The Importance of Failure," by John Unsworth, "talking about when we do these digital projects and things don't work out as planned, to not just chalk it up as a failure, but to write about it or talk about it and let others learn from the mistakes that were made—or the failure may not be the result of a mistake, but whatever led to the failure. And that also, that out of the failure always comes some new knowledge that's valuable" (professor, digital humanities).

A scientist who works on stream restoration noted that "in geomorphology, there is a tremendous acknowledgment of uncertainty and complexity, and so people don't generally say 'negative results.' They say, 'OK. Our hypothesis was this. The end result is this, which is different from what we thought, but that's still really useful to know.'" She also pointed out that the notion of failure "comes up in the geomorphology literature quite a bit . . . So you have people that are evaluating projects that were done by designers, by practitioners. You have academics writing about other people's failures, but you also have practitioners writing about their own failures. There has been a pretty big push by a lot of senior people in stream restoration to make it OK to talk about failure, and so people do talk about that. . . . So the big rationale is that stream restoration is a really new practice, and that there's an extremely steep learning curve. In many people's opinion we only just started figuring out how this works. There are a lot of people you could talk to, including me, who would say that, in fact, nobody in this country knows how to do stream restoration well, yet we spend billions of dollars on it every year. If we're going to get better, we have to be upfront about what's not working." She noted one senior researcher who had "been threatening for years to hold failure conferences, where all you do is get up and talk about failed projects" (assistant professor, geography).

Another individual in public administration research said she was pushing for more open discussion about errors and failures in her field. "I think it would be unusual in our field to admit to an error, so where we would use that term would be to offer proof to the reviewers and the readers that we have anticipated the kinds of errors that might happen and address them. . . . One of the things we have sort of pushed for within my field is a stronger admission by authors . . . about the limitations of their data and the limitations of their methodology. I'm in a very young field, so we're working so hard to generate knowledge. I think authors very often are not very willing to kind of really put their research of findings in some appropriate context. And so . . . in either my methodology or my conclusion in a manuscript I'd use 'error' to say, "Well, I anticipated sampling bias. I anticipated X or Y, and so to avoid those kinds of methodological errors I've done this" (professor, public and environmental affairs).

These interviewees describe, in a sense, the social dimension of what historian of science Friedrich Steinle (1997) described as "exploratory research." Those periods where no firm theoretical frameworks are in place are also character-ized by more open discussions about such strategies, and by greater tolerance for what appear to be unsuccessful trials. Researchers working in emerging fields are more liberal in their applications of standards for what is right or wrong, prom-ising or disappointing, aware that the standards themselves are still emerging. For other working scientists, by contrast, whatever is not going right in normal science is like a mistake: failure is experienced as a personal defeat, as an un-desirable, embarrassing, humiliating thing that seems to happen to nobody but oneself. The concept of failure is morally charged and is often associated with cheating and fraud.

Failure, Success, and Inconclusive Findings

In each instance of failure I have outlined so far, we have seen that what defines "failure" and "success" is local, contextual, shaped by the responses of others, and dependent upon contingent factors such as being able to convince one's peers of the advantages of a new way of doing science, finding a funding agency, or choosing a particular historiographical approach.

With regard to everyday science, however, the use of "failure" and "success" as analytic tools creates more problems than it solves. My discussion of instruc-tive mistakes implied that in everyday science, at least, successes and failures are easily recognizable by the practitioners themselves. However, in fact, recent ana-lyses suggest that even everyday situations are not completely determined by es-tablished research frameworks. Many outcomes are neither clear successes nor clear failures, and it is not always obvious how to proceed. Of course, in eve-ryday science, much of what scientists do is informed by an unstated and per-haps inexplicable knowledge of how to do things. But not all of everyday practice is guided and constrained in this sense. At many points in the research process, scientists have flexibility and options, and it is not always clear what decisions should be made. As Hans-Jörg Rheinberger put it in his discussion of biological experimentation, "Experimenters are usually working in a landscape where al-most nothing is either black or white, and almost everything consists of shades of gray. . . . [Experimental efforts] do not lead to promising findings, but they also do not lead to the clear-cut falsification of sharply delineated assumptions" (Rheinberger 2008, 76).

Notably, Rheinberger calls everyday practice "exploratory." His understanding of the term is different from the standard notion of "exploratory experimen-tation" as expounded by Steinle (1997). The latter notion was developed to ac-knowledge that there are periods in experimental science that are characterized by the absence of theoretical frameworks. Rheinberger emphasizes that even in situations where theoretical frameworks or paradigms are in place, everyday practice has the flexibility and openness that comes with the absence of sharply delineated assumptions.

Everyday science covers a spectrum from rule-guided or routine practice to very open, flexible procedures. When scientists find themselves uncertain and facing multiple options, they may have several ways to proceed. If they go one route and not another, they are not making a mistake. Rather, they are closing off certain plausible (perhaps even equally plausible) alternatives. The theoretical framework in place is not failing, the methodological strategies are not inadequate, and the tacit knowledge is not inapplicable; the background is just not specific enough to determine choices and procedures in everyday scientific practice. But if this is the case, then the conceptualization of scientific practice exclusively in terms of "failure" and "success" is misleading and may also have problematic implications, as would an account of scientific research that highlights only the successes. Some of the insights that philosophers of science gained in the analysis of theory change should be applied to periods of "normal" science as well, such as Kuhn's discussion about values in science (Kuhn 1977).

The interviewees were explicitly asked about inconclusive findings—what "inconclusive" meant to them, and whether they considered inconclusive findings significant enough to be communicated. Few researchers expanded on the topic. Only one person, a biologist, explicitly stated: "Many results are inconclusive." His explanation echoes the idea of "shades of gray" in experimentation: "You set up an experiment to test something, or to ask some question. You get your results, and you try to figure out what the results tell you, and the results may be very clear, very good signals in every aspect of the experiment, so they're positive results. But you can't infer anything significant from it. You didn't set up the experiment in a clean enough way to be able to get a conclusion. So that's inconclusive . . . Or you just don't have enough information to reach a valid conclusion. An alternate way of saying that would be that you've got too many alternative interpretations of the data. You can't choose among them, therefore that experiment is inconclusive. You need to add others to it to be able to distinguish among them" (professor, biology).

A physicist who made extensive comments about inconclusive results did not discuss findings *he* had found inconclusive. Rather, he described a scientific controversy in high-energy physics where different research groups disagreed about the existence of a particle. His story is really about replication attempts. "There was a discovery announced many years ago about a five-quark object that was observed by the Japanese at one of their accelerators. And because this was a specific new kind of particle that was being created, it showed up in their experiment with a very clear signature. . . . And what happened was, a number of other experiments looked, and some saw it. Some saw it in a different place, and some didn't see it at all. And then the discussion really got going [laughs] over what this meant, whether or not each experiment had some issue that would've generated a signal, and why is it that some of the better experiments with much higher statistics saw nothing at all. It was finally resolved in that people think the pentaquark no longer exists. . . . But this can lead to an incredibly complex situation sometimes, and that's one where multiple laboratories jumped on it as soon as the first observation was announced. . . . So it was a really open discussion for a while" (senior researcher, physics).

Notably, like failures, inconclusive results were also seen as a sign of incompetence. When asked about them, a cancer researcher said, "I think that's a lot of 'I'm not trying hard enough to figure out what's going on.' It's just a way to back out of something." The researcher added, "It drives me nuts. 'We had X, Y, and Z. We didn't see this, so it's kind of inconclusive.' Well, yeah, because you didn't ask the right questions. You didn't reframe your question, is what they do. A lot of people don't reframe their questions. I think that's the better way to put it" (associate professor, pediatrics). With the exception of the biologist, these researchers do not appear to think that "shades of gray" in experimentation are very common. Inconclusive results are stigmatized for the same reasons as things that do not work.

Failure *as* Success?

Today, scientists rarely address failures, errors, and research calamities in formal publications.[13] It is often assumed that current norms and standards of scholarly publication prevent scientists from sharing details about what was going on and going wrong in their research but that the scientists themselves *are* completely aware that failure and inconclusive findings are common occurrences. Firestein, writing for a broader readership, assumes that "it's general knowledge, inside of science." Similarly, Rheinberger (a molecular biologist turned historian of science) points out that it is "seldom acknowledged but certainly commonplace among working scientists that the bulk of their experimental efforts do not lead anywhere" (Rheinberger 2008, 76).

Interviews with scientists indicate that this perception is not commonplace. I can only draw on a limited number of interviews, but they suggest that researchers may not fully and consciously appreciate how common it is in scientific research that things don't work as expected, are inconclusive, or do not work at all, or that multiple avenues may open up at many points in their research. We saw that even though interviewees were explicitly invited to discuss error, failure, and inconclusive results, they rarely talked about how difficult it is to get an experiment to work and how frequently they failed, and they rarely reflected on the flexibility and openness of everyday research. They generally did not emphasize the toil and labor involved in getting experiments to run, the openness of the research process, and the myriad decisions that need to be made. Some explicitly said that "inconclusive" was not a term that came up in research contexts. It is as if the researchers assumed that inconclusive research outcomes and failures— mistakes—*ought not* to occur, which indicates that they define success in terms of reward.

In recent years, scientists have begun to use social media platforms to comment on their own and others' work.[14] On the face of it, the existence of hashtags such as #overlyhonestmethods and of blogs and other venues suggests that scientists are now more willing to share details about their research practice, are more tolerant of flexibility, and are more willing to recognize that much of their effort goes nowhere. However, a brief glance at tweets with the hashtag #overlyhonestmethods shows that sustained discussions of methodological issues are absent. Instead we find goofy comments, venting, and comic relief (see Bezuidenhout 2015). Yet it is too early to say what impact these new forms of communication will have. Perhaps failure will eventually be less stigmatized. Some commentators, however, have already suggested that the tone of online criticism of others' research has actually become more toxic and hostile than in the past, so much so that *Nature* editorials have called for more courtesy and humility in online forums ("Post-Publication Criticism" 2016). While more information about specific research practices becomes available, the soundbites on social media appear to be heavily emotionalized.

How scientists think about failure and about scientific practice more generally matters. Several common ideas about science appear to be genuine obstacles for effective peer criticism and open discussion of methodological issues in established fields. These include the idea that everyday science is "day science" and takes place within a framework of well-defined, agreed-upon procedures; the close conceptual and psychological association of failure with personal defeat, and with sloppy or even questionable research practices; and the conflation of failure and negative results. One of the lessons from my discussion of failure and success is therefore a new agenda for science educators. Instructors need to address with their students both the openness of everyday science and the double misconception of failure as personal defeat. Scientists need to have a higher tolerance for and awareness of uncertainty, "shades of gray" in everyday contexts, and realistic expectations of day-to-day activities.

Here is a role for historical analysis in science education. Historical examples of the practice of research can help spread the knowledge that everyday science is not as firmly rule-guided as one might think, and that uncertainty and failing are rather common in everyday science. But not any historical analysis will do. Presenting episodes from the past (including the recent past) to science students is beneficial if we pay attention to all of the details of scientific toil and labor, to the flexibility and the many dead ends, detours, and failed trials. Yet recasting failures *as* successes creates more confusion than it removes.

Notes

* This chapter originates from a paper I presented at the workshop "Science, Technology and the Good Life" at the University of Notre Dame, April 2018. I thank the organizers for inviting me to the conference, for the invitation to contribute to this volume, and for their editorial suggestions. I am grateful to the workshop participants for their feedback on my presentation, to Dori Beeler and Ann-Sophie Barwich for their helpful comments on an earlier version of this chapter, and especially to Jordi Cat for his incisive critique of the penultimate draft. I learned a lot at "The Success of Failure: Perspectives from the Arts, Sciences, Humanities, Education, and Law," a conference at Columbia University in December 2017, and I thank Stuart Firestein for inviting me to participate in it. I also wish to acknowledge my former collaborators Nora Hangel, with whom I had several discussions about the challenges of analyzing interviews with scientists about their research practices, and Cassidy Sugimoto, with whom I conducted the interviews I discuss in this chapter.

1. The survey and interviews were carried out in cooperation with Dr. Cassidy Sugimoto, School of Informatics, Computing, and Engineering, Indiana University, Bloomington.

2. Interviewees were identified through an online survey designed to examine scientists' experiences with unpublishable research (see Tsou et al. 2014 for details and results). The last question on the survey queried the respondents' willingness to be contacted for a follow-up interview. Twenty-eight percent of respondents (N = 125) indicated that they were willing to be contacted for this purpose. Ultimately, forty individuals were interviewed. All interviews were recorded except one; another could not be transcribed due to background noise. The interviews focused on conceptualizing "unpublishable" units, the degree to which certain types of unpublishables existed within domains, experience in publishing these types of papers, and the perceived value of such work.

3. For a similar approach, see Hangel and Schickore 2017; Schickore and Hangel 2019.

4. The list comprised the terms "negative result," "null result," "inconclusive result," "error," "artifact," "failure," and "replication."

5. The first quote is ascribed to Henry Ford. For the second quote, variations on it, and possible sources, see https://quoteinvestigator.com/2014/12/16/no-mistakes/.

6. Firestein (2015, 71–95) discusses the importance of "teaching failure."

7. Notably, such stories of struggle stand in contrast to the message that informal science education is giving: science museums, zoos, and similar institutions make science *fun* and seek to recruit young people through entertainment.

8. A similar argument could also be made with regard to some personal failures: we could imagine that progress in a particular field became possible as a researcher whose

approach dominated a field declined in strength or activity, and because that person's influence waned, alternative approaches could flourish.

9. "All of the theatre works I discuss can be considered 'experimental' in the broad generic sense . . . , meaning that they play with and examine the limits of (theatre) form, putting pressure on certain conventions and prescribed rules in order to stretch or invent others" (Bailes 2010, 13).

10. When he published this book, Morton Meyers, a medical doctor, had retired from his university position.

11. Kuhn noted that for paradigm change to happen, advocates of the old paradigm would have to die out.

12. Firestein's notion of iconoclastic failure does not apply to emerging research fields. Because the concepts and rules of emerging fields are still in flux, there is no point in explicitly defying them. Yet those who enter emerging fields may be perceived as iconoclasts by the members of established fields.

13. Before 1900, scientists were quite willing to admit that, most of the time, things did not work, or at least did not work as expected. See Schickore 2017 for more detailed accounts of writing procedures in experimental reports prior to 1900.

14. I am grateful to Ann-Sophie Barwich for drawing my attention to #overlyhonest methods.

References

Athanassoulis, N. 2017. "A Positive Role for Failure in Virtue Education." *Journal of Moral Education* 46, no. 4: 347–62.

Babur, O. 2018. "Talking About Failure Is Crucial for Growth." *New York Times*, August 19.

Bailes, S. J. 2010. *Performance Theatre and the Poetics of Failure*. New York: Routledge.

Bezuidenhout, L. 2015. "Variations in Scientific Data Production: What Can We Learn from #overlyhonestmethods?" *Science and Engineering Ethics* 21: 1509–23.

Collins, H. 2017. *Gravity's Kiss*. Cambridge, MA: MIT Press.

Firestein, S. 2015. *Failure: Why Science Is So Successful*. Oxford: Oxford University Press.

Galison, P. 1987. *How Experiments End*. Chicago: University of Chicago Press.

Hangel, N., and J. Schickore. 2017. "Scientists' Conceptions of Good Research Practice." *Perspectives on Science* 25: 766–91.

Hon, G. 1995. "Going Wrong: To Make a Mistake, to Fall into an Error." *Review of Metaphysics* 49: 3–20.

Jacob, F. 1998. *Of Flies, Mice, and Men*. Cambridge, MA: Harvard University Press.

Krimsky, S. 2003. *Science in the Private Interest*. Lanham, MD: Rowman & Littlefield.

Kuhn, T. S. 1977. "Objectivity, Value Judgment, and Theory Choice." In *The Essential Tension*, 320–39. Chicago: University of Chicago Press.

Lin-Siegler, X., et al. 2016. "Even Einstein Struggled: Effects of Learning About Great Scientists' Struggles on High School Students' Motivation to Learn Science." *Journal of Educational Psychology* 108: 314–28.

Mach, E. 1976. *Knowledge and Error*. Dordrecht: Reidel.

Meyers, M. 2007. *Happy Accidents: Serendipity in Modern Medical Breakthroughs*. New York: Arcade.

Mulkay, M., and N. Gilbert. 1982. "Accounting for Error: How Scientists Construct Their Social World When They Account for Correct and Incorrect Belief." *Sociology* 16: 165–83.

Nickelsen, K., and G. Graßhoff. 2008. "Concepts from the Bench: Hans Krebs, Kurt Henseleit and the Urea Cycle." In *Going Amiss in Experimental Research*, edited by G. Hon, J. Schickore, and F. Steinle, 91–117. Dordrecht: Springer.

Parkes, E. 2019. "Scientific Progress Is Built on Failure." *Nature Research*, January 10. DOI: 10.1038/d41586-019-00107-y.

"Post-Publication Criticism Is Crucial, but Should Be Constructive." 2016. [Editorial.] *Nature* 540 (7631): 7–8.

Rheinberger, H.-J. 2008. "Experimental Reorientations." In *Going Amiss in Experimental Research*, edited by G. Hon, J. Schickore, and F. Steinle, 75–90. Dordrecht: Springer.

Schickore, J. 2017. *About Method: Experimenters, Snake Venom, and the History of Writing Scientifically*. Chicago: University of Chicago Press.

Schickore, J., and N. Hangel. 2019. "'It Might Be This, It Should Be That . . .': Uncertainty and Doubt in Day-to-Day Research Practice." *European Journal of Philosophy of Science* 9: 1–21.

Sismondo, S. 2009. "Ghosts in the Machine: Publication Planning in the Medical Sciences." *Social Studies of Science* 39: 171–98.

Steinle, F. 1997. "Entering New Fields: Exploratory Uses of Experimentation." *Philosophy of Science Supplement* 64: S65–S74.

Steinle, F. 2016. *Exploratory Experiments: Ampère, Faraday, and the Origins of Electrodynamics*. Pittsburgh: University of Pittsburgh Press.

Tsou, A., et al. 2014. "Unpublishable Research: Examining and Organizing the 'File Drawer.'" *Learned Publishing* 27: 253–58.

Vallery, T. 2016. "Growth from Failure." *Science* 353 (6298): 514.

Zaringhalam, M. 2016. "Failure in Science Is Frequent and Inevitable—and We Should Talk More About It." Guest blog, *Scientific American*, June 30. https://blogs.scientificamerican.com/guest-blog/failure-in-science-is-frequent-and-inevitable-and-we-should-talk-more-about-it/.

11

Virtues in Scientific Practice*

Dana Tulodziecki

Introduction

One of the main venues for the discussion of theoretical or epistemic virtues in philosophy of science has been the debate about scientific realism. Scientific realists believe that mature scientific theories are approximately true, including what they say about entities, processes, and mechanisms that we cannot directly observe. Anti-realists deny this. They agree with the realist that claims about such unobservables have truth-values, and they even agree that science might attain such truths—but what they vehemently deny is that we can ever be *justified* in believing such claims. One important argument that anti-realists have put forward to argue for this conclusion is the argument from underdetermination. Informally, the argument goes like this:

1. The available observational evidence (including all *possible* future evidence) always supports two (or more) theories that cannot both be true.
2. Our only reason for believing our scientific theories to be true is the observable evidence on which they are based.
3. Hence, we have no reason to prefer any one of these theories to any other.

According to the anti-realist, this demonstrates that there are no reasons (and, in fact, that there cannot be any such reasons) to accept those parts of a theory that transcend the data and, since this covers all scientific theories, the anti-realist concludes that we cannot have any epistemic reasons to prefer one scientific theory to others, at least as long as the observational data for the theories in question is the same.[1]

One of the ways in which realists try to counter this argument is by denying its second premise: they argue that, in addition to the empirical evidence, there are other epistemic criteria that can help us choose one particular theory over others. Thus, according to the realist, while it might be true that there are empirically equivalent and logically incompatible rival theories, such theories are not epistemically equivalent, because they are unequal with respect to some other factor relevant for theory choice. The most popular candidates for such criteria

Dana Tulodziecki, *Virtues in Scientific Practice* In: *Science, Technology, and Virtues*. Edited by: Emanuele Ratti and Thomas A. Stapleford, Oxford University Press. © Oxford University Press 2021. DOI: 10.1093/oso/9780190081713.003.0012 have been the so-called theoretical virtues, properties of our scientific theories such as coherence with other (established) theories, unifying power, consilience, generation of novel predictions, explanatory power, simplicity, elegance, parsimony, lack of ad hoc features, or fruitfulness. While realists think that (some of) these virtues have epistemic power, anti-realists think they are merely pragmatic, believing that even if these factors play a role in theory choice, they fail to be epistemically significant. In this vein, van Fraassen, for example, explicitly claims that the theoretical virtues "cannot rationally guide our epistemic attitudes and decisions" (1980, 87).

Thus, as we can see, part of the question that is at stake in the debate about underdetermination and the virtues is the more general question of what constitutes an epistemically relevant factor for scientific theory choice, with anti-realists claiming that only the empirical evidence (or entailment thereof) counts, while realists hold either that the theoretical virtues themselves are somehow evidential or that, at the very least, they are capable of breaking the alleged epistemic tie between the rival theories in question. However, despite the fact that the virtues frequently feature in realist responses to the underdetermination argument, there have not been any real attempts to develop substantial accounts of the virtues that could make them function robustly either in debates about the argument itself or in debates about the dimensions of justification of our scientific theories more generally.[2] There are some accounts of individual virtues;[3] however, these accounts focus on specific virtues and do not aim to be more general accounts of how epistemic virtues function in science, nor do they address questions about how we might think about the epistemic function of various theoretical properties as they feature more generally in contexts of scientific justification.[4] Moreover, many of the debates about the virtues in theory choice have taken place on an abstract level, often invoking algorithms or skeptical hypotheses, and emphasizing the "in principle nature" of arguments for and against the virtues (Psillos 1999; Kukla 1993, 1996).

In this chapter, I want to relocate the debate about the virtues to the empirical level and argue that the question of whether the virtues (and what virtues, if any) have epistemic import is best answered empirically, through an examination of actual scientific theories and hypotheses in the history of science. As a concrete example of how this approach works, I will discuss in some detail a case study from the history of medicine, concerning the mid-nineteenth-century debate about the transmissibility of puerperal fever. I will show how some specific virtues were put to work in this particular case, and argue that an analysis of this episode suggests that the virtues are at least sometimes epistemic. Then, using the case study as a basis, I will explain what is required in order to make a more general argument for the epistemic potential of the theoretical virtues along these lines. Part of my aim in this chapter is to show that it is possible, at least in

principle, to uncover both theoretical virtues and their role in scientific justification through engaging in such detailed case studies. Whether these virtues are
ultimately to be regarded as generally epistemic or not is, at this point, an open
question; however, the main point of this chapter is not to show that specific
virtues, or even the virtues in general, *are* epistemic in nature (although I think
its conclusions are suggestive of such a view) but, rather, to make plausible the
view that the virtue question is one that can be settled empirically, and to show
that answers about the epistemic standing of the virtues will involve a robust engagement with contexts in which we see these virtues at play. Moreover, thinking
about the virtue question as an empirical one is an important beginning for a
more systematic and extensive account of how to discover and put to use factors
other than empirical evidence in our thinking about the epistemic ingredients
and justification of our scientific theories.

Further, I will argue that despite the fact that the case shows that the virtues
are at least sometimes epistemic, this alone does not support the realist's view
that the virtues are an indicator of a theory's (approximate) truth. Neither, however, does it support the anti-realist's view that the virtues are merely pragmatic.
Instead, an examination of the case suggests that the way the virtue question
is framed in the scientific realism debate is not helpful in uncovering the ways
in which the virtues work in actual cases of scientific practice. This is because
this debate presupposes a misleading dichotomy of the virtues as either truth-
conducive or non-epistemic, when an examination of actual cases suggests that
it is more appropriate to view the status of the virtues as falling somewhere in
between these two categories.

I will proceed as follows. First the chapter focuses on some competing claims
from the late eighteenth and early to mid-nineteenth centuries concerning the
transmissibility of puerperal fever, showing how it is, at least in principle, possible to link the theoretical virtues to empirical data. Then I argue that this case
shows that the virtues are least sometimes epistemic, but also that neither realists
nor anti-realists get it quite right: the virtues, even if epistemic, are not necessarily truth-conducive, but neither are they merely pragmatic. I also argue that
the discussion of puerperal fever shows that the virtue question, as it is currently
featured in the realism debate, ought to be reformulated. We should examine not
just whether a given scientific theory has virtues or not but, rather, how debates
among competing theories, all of which have some virtues, get resolved.

Puerperal Fever in the Mid-1800s

If the virtue question is to be thought of as an empirical one, the question arises
as to what sort of empirical data is required to shed light on the virtues. My goal

in this section is to provide an example of this sort of data, both to show in more detail what form it might take and also to show that it is, in fact, possible to obtain it. Specifically, I want to show that it is possible to obtain it through the type of case study that I am offering here (and, by extension, through others like it). One case study by itself cannot establish the epistemic importance of anything; however, it does suggest that the virtues can function in an epistemic role, by showing just how they were made use of by the relevant practitioners at the time, who appealed to them as considerations in favor of their hypotheses over others.

The case study I look at is that of puerperal fever in the mid-1800s. This case was made famous in philosophy by Hempel's discussion of Semmelweis in *Philosophy of Natural Science* (1966), and it has been taken up by a number of philosophers since (Lipton [1991] 2004; Gillies 2005; Bird 2007, 2010; Scholl 2013). Here, however, I am interested not so much in the various controversies surrounding Semmelweis (for a discussion of these, see Tulodziecki 2013b) as in an examination of the different hypotheses that were put forward with respect to puerperal fever, and with the sorts of considerations that were adduced in their favor at the time.

Puerperal fever, a bacterial infection of the reproductive tract following childbirth, was the most common cause of maternal mortality in the nineteenth century. Appearing after a woman had given birth, its symptoms included shivering, a high pulse, fever, and extraordinary abdominal pain (to the extent that even being covered with sheets or blankets could not be tolerated). The sporadic version had a mortality rate of about 35 percent; in epidemics, the mortality rate was as high as 80 percent.[5] The medical situation with respect to puerperal fever (and most other diseases, for that matter) was rather complicated: there was no consensus on virtually any aspect of the disease—its symptoms, its pathology, whether slightly different sets of symptoms ought to all be classified as puerperal fever or as different diseases, whether there was one type or many, whether it was transmissible from person to person (and, if so, in what ways), whether it had one cause or many, what the right treatments were, whether different treatments were called for under different circumstances, what factors would exacerbate the prevalence of the disease, and who was particularly vulnerable to it. The only thing there was agreement on was that it caused a lot of deaths.

The dominant disease theory of puerperal fever at the time was the miasma theory, according to which puerperal fever was caused by toxic odors that resulted from the decomposition of organic matter. Most of these miasmas were atmospheric, but there were also thought to be telluric influences rising from the ground. Because these were thought to play a crucial role in the causal origins of puerperal fever, emphasis was put, for example, on the locations of hospitals, to ensure that the buildings would not be close to swamps, that they were sufficiently elevated, and so on. There were a number of other factors that were thought to be relevant as well, such as overcrowding and bad ventilation. In addition, it was thought that people of certain predispositions—for example, women who had a "bad constitution" or had gone through a difficult labor—were particularly prone to falling ill. While these other factors were thought to play a role, by themselves they were not taken to be sufficient for causing the disease; some miasmatic or telluric influence or other was generally thought to be required for doing so.[6]

As we have seen, there were questions about virtually every aspect of puerperal fever. Since it is impossible to do justice to all of these issues, in what follows I will focus only on one specific question: that of the communicability or transmissibility of the disease. Proponents of the transmissibility hypothesis believed that puerperal fever was communicable—that is, previous cases of the disease could cause later cases of the disease, and that it could be communicated, either directly from one patient to the next, or indirectly "through the medium of a third person; and that person generally the medical attendant or nurse" (Simpson [1851] 1871, 507). Opponents of the transmissibility hypothesis rejected the idea that puerperal fever was transmissible and believed that it was always caused solely by a combination of atmospheric and telluric influences, in conjunction with the victim's predispositions.

What sorts of considerations did proponents of the transmissibility hypothesis appeal to? Why did they think it was possible for puerperal fever to be transmitted via doctors and midwives? In what follows, I will discuss three theoretical virtues that are prominent on realists' lists—explanatory power, consilience, and the generation of novel predictions—and show that these were indeed considerations that played an important and, moreover, epistemic role in the arguments for the transmissibility hypothesis.[7]

Explanatory Power

One consideration consistently appealed to by virtually every proponent of the transmissibility hypothesis was that this hypothesis could explain a number of phenomena the non-communicability hypothesis could not. For example, it was pointed out that puerperal fever often followed a single practitioner "with the keenness of a beagle" (Holmes 1892, 157). Here is a typical passage from James Young Simpson, a mid-nineteenth-century Scottish physician (also famous for introducing chloroform into anesthesia during birth), describing this phenomenon, which was a frequent occurrence:

> [T]hat it [puerperal fever] was so propagated by the medical attendant or nurse, we further believe upon the following species of evidence –viz. that it

was . . . distinctly and precisely limited to the practice of one or two practitioners only, out of a large number of medical practitioners, practicing in a large community. Many examples were recorded, and many more unrecorded were known to the profession, of the disease being this limited to the practice of a single practitioner in a town or city; all, or almost all, the patients of that practitioner being affected with it, where none of the patients of other practitioners were seized with any attack of the disease. In these cases we could not believe it to be owing to any morbific influence present in the air, or emanating from the locality in these cities or towns. For if so, it would affect indiscriminately the patients of all practitioners. But it had been often seen, as it was just now remarked, to haunt the steps of a single practitioner, and a single practitioner only, in a community. (Simpson [1851] 1871, 507–8)

The fact that the transmissibility hypothesis can explain what the non-communicability hypothesis cannot is considered to be "a species of evidence" for the transmissibility view. Simpson points out that the non-communicability hypothesis simply cannot account for the existing patterns of infection, while the transmissibility hypothesis can. If the non-communicability hypothesis had been correct, one should have found a more uniform pattern of infection in areas similar to each other—for example, most women in a certain area ought to have been affected by the disease, not just a small number of specific individuals. Proponents of this view might have tried to account for this pattern by appealing to a combination of other factors (poor constitution, difficult labor, etc.). However, even in that case, there was no reason for the link between the disease and specific doctors or midwives, such as that referenced in the following passage, an example of a common type of story:

Dr. Roberton, of Manchester, tells us, that in 1840 upwards of 400 women were delivered by different midwives in connection with the Lying-in Hospital in Manchester. These 400 women were delivered in different parts of the town at their own houses: 16 of them died of puerperal fever; all the others made good recoveries. The production of this could not have arisen from any general epidemic, or atmospheric or telluric influence; for the fatal cases occurred in no one particular district, but were scattered through different parts of the town. Now, these 400 women and more were attended in their confinements by twelve different midwives. Eleven of these twelve midwives had no puerperal fever amongst their patients. The sixteen fatal cases had occurred in the practice of one only of the twelve. The disease, in fact, was limited entirely to her patients. There must have been something, then, connected with that one midwife, in which she differed from the other midwives, inasmuch as all her patients took the disease, whilst the patients of all the other midwives escaped from it. And in

medical philosophy we cannot fancy that this something consisted of aught else
than some form of that morbific principle or virus to which pathologists give
the name of contagion. (Simpson [1851] 1871, 508)

The non-communicability hypothesis, even one taking into account a number
of modifying factors, such as individual predispositions and circumstances,
had difficulty accounting for why the disease followed specific physicians and
midwives. If this phenomenon had been an isolated one, one might have put it
down to coincidence—a practitioner might have been unlucky in happening
to attend women with particularly weak constitutions, for example—but the
point being made is precisely that this phenomenon was widespread and the
norm rather than the exception, and that *that* is something for which the non-
communicability hypothesis cannot provide an explanation. Holmes, for example, is clear about what the issue is:

> The question always comes to this,—Is the circumstance of intercourse with the
> sick followed by the appearance of the disease in a proportion of cases so much
> greater than any other circumstance common to any portion of the inhabitants
> of the place under observation, as to make it inconceivable that the succession
> of cases occurring in persons having that intercourse should have been the re-
> sult of chance? If so, the inference is unavoidable, that the intercourse must
> have acted as a cause of the disease. (1855, 17)

And, he thinks, the evidence makes clear that this inference is indeed
unavoidable:

> I have had the chances calculated by a competent person, that a given practi-
> tioner, A., shall have sixteen fatal cases in a month, on the following data: A. to
> average attendance upon two hundred and fifty births in a year; three deaths in
> one thousand births to be assumed as the average from puerperal fever; no ep-
> idemic to be at the time prevailing. It follows, from the answer given me, that if
> we suppose every one of the five hundred thousand annual births of England to
> have been recorded during the last half century, there would not be one chance
> in a million million million millions, that one such series should be noted. No
> possible fractional error in this calculation can render the chance a working
> probability. Applied to dozens of series of various lengths, it is obviously an
> absurdity. Chance, therefore, is out of the question as an explanation of the
> admitted coincidences. (1855, 13)

He concludes: "There is, therefore, *some* relation of cause and effect, between the
physician's presence and the patient's disease" (1855, 13).

There were also additional patterns that the non-communicability hypothesis lacked any sort of account of and that had a perfectly good explanation according to the transmissibility hypothesis, according to which such patterns were to be expected. These additional patterns were constituted, not by cases in which the disease followed a particular individual, such as above, but by cases in which it was possible to trace "paths of infection": a practitioner gave the disease to certain women, who passed it along to other practitioners who treated them and who in turn spread it among their patients and yet other practitioners who treated those patients. A striking example of this sort of data comes from Alexander Gordon as early as 1795 (see Figure 11.1):

> The midwife who delivered No. 1 in the table [Fig. 11.1], carried the infection to No. 2, the next woman whom she delivered. The physician who attended Nos. 1 and 2, carried the infection to Nos. 5 and 6, who were delivered by him,

A TABLE,—Containing an account of those Patients affected with the Puerperal Fever, who were attended by Dr. Gordon, from December 1789 to October 1792.

When taken ill.	No.	Name.	Age.	Residence.	Cured.	Dead.	By whom delivered.
1789.							
December	1	James Garrow's wife	27	Woolman-hill		5th day	Mrs. Blake.
Ditto	2	James Smith's wife	30	Ditto		23d „	Ditto.
Ditto	3	John Smith's wife	34	Green		11th „	Mrs. Elgin.
Ditto	4	Al. Mennie's wife	25	Hardgate		11th „	Ditto.
1790.							
January	5	John Anthony's wife	25	North-street		3d „	Dr. Gordon.
February	6	Christian Durward	36	Rottenholes		3d „	Ditto.
April	7	Al. Stuart's wife	30	Denburn	1		Mrs. Philp.
May	8	William Elrick's wife	34	Exchequer-wynd	2		Mrs. Blake.
Ditto	9	Elizabeth Murray	28	North-street		7th „	Ditto.
Ditto	10	Helen Mitchell	30	Ditto	3		Ditto.
Ditto	11	Janet Wier	34	Denburn	4		Mrs. Elgin.
August	12	Mrs. Johnston	36	Littlejohn's-st.	5		Mrs. Smith.
Ditto	13	Geo. Webster's wife	38	Fowler's-wynd	6		Mrs. Blake.
Ditto	14	Peter Paul's wife	32	Windmill-brae	7		Ditto.
Ditto	15	John Low's wife	25	Justice-mills		5th „	Mrs. Smith.
Ditto	16	Mrs. Milne	27	North-street	8		Mrs. Blake.
Septemb.	17	Isabel Allan	36	Birnie's-close		5th „	Mrs. Coutts.
Ditto	18	Robert Burr's wife	30	Gallowgate		2d „	Mrs. Irvine.
October	19	Al. Eddy's wife	36	Ditto		3d „	Mrs. Clark.
Ditto	20	Agnes Milne	24	Putachie-side	9		Ditto.
Ditto	21	Al. Stuart's wife	26	Green	10		Mrs. Blake.
Ditto	22	Elizabeth Jamieson	25	Windmill-brae		5th „	Dr. Gordon.
Ditto	23	Dundas Nicol's wife	25	Green	11		Mrs. Philp.
Ditto	24	Al. Brown's wife	27	Loan-head		5th „	Mrs. Elgin.

Figure 11.1 Gordon's "Paths of Infection"

Source: Alexander Gordon, "Containing an Account of Those Patients Affected with the Puerperal Fever, Who Were Attended by Dr. Gordon, from December 1789 to October 1792," in *Essays on the Puerperal Fever and Other Diseases Peculiar to Women*, ed. F. Churchill (London: Sydenham Society, 1849), 452–453, table.

and to many others. The midwife who delivered No. 3 carried the infection to No. 4, from No. 24 to Nos. 25, 26, and successively to every woman whom she delivered. The same thing is true of many others, too tedious to be enumerated. (Gordon [1795] 1849, 471)

Whereas before it had remained mysterious why particular women would contract the disease, the transmissibility hypothesis could explain who got infected and how. The only option for the non-communicability hypothesis was, once again, to appeal to additional criteria besides the atmospheric-telluric influence. But even if such a story could be told, the different cases above still would have required separate and different explanations, whereas the transmissibility hypothesis could supply a unified explanation for all cases in one stroke. Moreover, in the same way, it could also account for new and previously puzzling epidemics: "The midwife who delivered Mrs. K—— carried the infection to No. 55 in Nigg, a country parish not far from Aberdeen, from whom it spread through the whole parish" (Gordon [1795] 1849, 471).

Lastly, the transmissibility hypothesis could explain the existence of anomalous regions that, mysteriously, were free from the disease. On the non-communicability hypothesis, there was no explanation for the existence of these regions; after all, if atmospheric-telluric conditions were responsible, they should have affected areas in their entirety, without pockets free from the disease. Thus, for example, it seemed "remarkable, that the puerperal fever should prevail in the new town, and not in the old town of Aberdeen, which is only a mile distant from the former" (Gordon [1795] 1849, 472). Once again, however, Gordon has an explanation: "[T]he mystery is explained when I inform the reader, that the midwife, Mrs. Jeffries, who had all the practice of that town, was so very fortunate as not to fall in with the infection, otherwise the women, who she delivered, would have shared the fate of others." He concludes that all this shows that "the cause of the puerperal fever, of which I treat, was a special contagion or infection, altogether unconnected with a noxious constitution of the atmosphere" ([1795] 1849, 472).

To sum up: the fact that the transmissibility hypothesis could explain these various patterns and that the non-communicability hypothesis could not was taken to count in favor of the transmissibility hypothesis. Among the phenomena that lacked an explanation on the non-communicability view but were explained and, indeed, expected on the transmissibility view were (1) the fact that the disease tended to follow specific practitioners, (2) that it was possible to trace specific and detailed paths of infection both locally and across geographical areas, (3) that it could explain why exactly those people who fell sick fell sick, (4) how new epidemics came about, (5) why there existed anomalous regions free of the disease, and (6) how, precisely, the disease was spread through different parts of town, different towns, and different geographical regions.

Consilience

Another much-cited virtue is consilience. This virtue goes back to Whewell, who explains it in the following way:

> [T]he evidence in favour of our induction is of a much higher and more forcible character when it enables us to explain and determine [i.e., predict] cases of a kind different from those which were contemplated in the formation of our hypothesis. The instances in which this has occurred, indeed, impress us with a conviction that the truth of our hypothesis is certain. (1858, 87–88)

As we can see, the idea behind this virtue is that it speaks in favor of a hypothesis if that hypothesis can account for types of phenomena that did not play a role in the formation of the original hypothesis. It turns out that this is the case for the transmissibility hypothesis: the hypothesis that puerperal fever is transmissible through medical practitioners can explain a number of phenomena "different from those which were contemplated in the formation of our hypothesis." For example, the transmissibility hypothesis can explain a variety of phenomena concerning erysipelas (a streptococcal rash) even though considerations about erysipelas were not part of the original evidence for the transmissibility hypothesis. Among the claims being made by proponents of the transmissibility hypotheses was

> that when the fingers of medical men were impregnated with the morbid secretions thrown out in erysipelatous inflammation, the inoculation of these matters into the genital canals of parturient females produced puerperal fever in them in the same way as the inoculation of the secretions from patients who had died of puerperal fever itself. The effused morbid matters in the one disease, as in the other, were capable of producing the same effect when introduced into the vagina of a puerperal patient. (Simpson [1851] 1871, 516)

As before, we have plenty of stories being invoked to support this conclusion. Here is a representative example:

> In an instance recorded by Mr. Hutchinson, two surgeons, living at ten miles' distance from each other, met half-way to make incisions into a limb affected with erysipelas and sloughing. Both practitioners touched and handled the inflamed and sloughing parts; and the first parturient patients that both practitioners attended within thirty of forty hours afterwards, in their own distant but respective localities, were attacked with, and died of, puerperal fever. The late Mr. Ingleby mentions an instance of a practitioner making incisions

into structures affected with erysipelas, and going directly from this patient to a patient in labour. This patient took puerperal fever and died. And within the course of the next two days, seven cases of puerperal fever occurred in the practice of the same practitioner, almost all of them proving fatal. And various other cases, similar to the preceding, were well known to the profession. (Simpson [1851] 1871, 516)

In addition, it was observed that the connection between puerperal fever and erysipelas went both ways: "Not only was the morbid matter in erysipelas apparently sometimes capable of producing puerperal fever, but the secretions and exhalations from puerperal fever patients seemed, on the other hand, sometimes capable of producing erysipelas" (Simpson [1851] 1871, 516–17). In fact, it was pointed out not just that erysipelas produced puerperal fever in patients of doctors who had treated erysipelas but, moreover, that the patients' secretions "produced also erysipelas in several of the nurses, relations, and attendants upon the patients" (Simpson [1851] 1871, 516–17). For example, a doctor named Sidey had a patient die of puerperal fever, and in the week following the patient's death,

the patient's mother-in-law, who was in constant attendance upon her, was attacked with fever and erysipelas of the face and head. One of the patient's sons, a boy five years of age, was attacked with erysipelas of the face; a daughter was seized with fever and sore throat, with dusky redness, which continued for some time; and the patient's sister-in-law was attacked with acute gastric symptoms, and great abdominal irritation, under which she sank in a few days. Here we have apparently the same focus on contagion producing puerperal fever in puerperal patients, and erysipelas, inflammatory sore throat, etc., in patients who were not in a puerperal state. (Simpson [1851] 1871, 516–17)

Lastly, it was pointed out that "the two diseases had in Britain been repeatedly observed to prevail at the same time, in the same town, in the same hospital, or even in the same wards" (Simpson [1851] 1871, 515–16). All these phenomena would have been puzzling on the assumption that the non-communicability hypothesis was true, yet they were exactly what was to be expected on the transmissibility hypothesis. Moreover, the original formulation of the transmissibility hypothesis was based not on evidence about erysipelas but, rather, on phenomena concerning the connection between the onset of puerperal fever and treatment by specific individuals who had previously been associated with the disease in one way or another. Yet the transmissibility hypothesis could "explain and determine" these different types of cases, specifically (1) the observation that puerperal fever was often contracted by patients whose doctors had previously treated cases of erysipelas, (2) the observation that cases of erysipelas often followed incidents of childbed fever, and (3) the coinciding of cases and epidemics of erysipelas and puerperal fever.

The Generation of Novel Predictions

To end, I want to briefly draw attention to one last virtue that is to be found on virtually every list: the generation of novel predictions. Again, this is something we can clearly see exemplified by the transmissibility hypothesis and not by the non-communicability hypothesis: while the non-communicability hypothesis was able to tell, in certain cases, a story of how the disease might have come about in certain areas (atmospheric-telluric conditions) and in certain people (a combination of conditions and predispositions), it failed to provide any sort of systematic account of why particular individuals fell sick. With hindsight, it was possible for the theory to look at a sick person and invoke various factors that might have contributed to that person's falling sick; however, the reasons were different every time, and while it was always possible to appeal to certain contributing factors (say, a weak constitution), even this could not be done systematically, since it just wasn't the case that there was an actual correspondence between these properties and incidences of the disease. For example, it simply was not the case that, generally speaking, people with sickly dispositions, those who had gone through a difficult labor, or those who had conceived out of wedlock were more prone to the disease, even though these factors would with hindsight be invoked as explanations for why specific women fell ill. Indeed, the list of potential factors was so long that some explanation along the above lines could always be found, since the number of different factors and combinations of factors was huge. Despite this, however, the non-communicability hypothesis was in no position to make any predictions whatsoever: it could not predict who would fall ill or what groups of people might fall ill, since no cause by itself was deemed sufficient, and any combination of factors might or might not actually bring about the disease. The case was quite different for the transmissibility hypothesis. It could predict, quite neatly, that if a certain practitioner had been in touch with victims of childbed fever, there would likely be death among his or her patients in the immediate future; it could predict that "thorough cleansing" on the practitioner's part would diminish this possibility; it could predict that if there were cases of erysipelas, childbed fever would soon follow (and the other way around), that outbreaks of surgical fever in hospitals would usually be followed by outbreaks of puerperal fever in the nearby maternity wards (due to cross-contamination and lack of hygienic measures on the part of surgeons who would move frequently between the maternity wards and operating theaters), and so on. In short, it made predictions about who the likely next victims would be. As Gordon puts it:

I could venture to foretell what women would be affected with the disease, upon hearing by what midwife they were to be delivered, or by what nurse they were to be attended during their lying-in; and almost in every instance my prediction was verified. ([1795] 1849, 447)

So, as we can see, in the case of the debate about the origin of puerperal fever, some of the most prominent theoretical virtues—explanatory power, consilience, and generation of novel predictions—were invoked in favor of the transmissibility hypothesis, while they do not feature similarly in the non-communicability hypothesis.

One natural conclusion from the way the virtues play into the debate about the transmissibility of puerperal fever seems to be that the winning hypothesis (that puerperal fever was communicable) exemplified the virtues, whereas the losing hypothesis (that puerperal fever was not communicable) did not. Moreover, it seems easy, in retrospect, to regard the transmissibility hypothesis as an early version of modern contagionist views, maybe even a precursor to the germ theory, and to identify the non-communicability view with an outdated anti-contagionist perspective. Passages such as the following make this particularly easy for us. Simpson, for example, writes that he had been

informed of an instance by Professor Patterson, in which a medical gentleman, after having lost several cases of puerperal fever, got rid of the disease in his practice by changing his clothes, and using chloride of lime, etc.; but it again returned to him when he happened to deliver a patient immediately after wearing a pair of gloves which he had used during the time of the puerperal epidemic; and certainly, if there was any piece of dress more apt to retain the contagion than another, it was this useless and superfluous appendage to our attire; for it might retain the morbid secretions that were originally on the fingers of the accoucheur, just as our vaccinating glasses would retain the cow-pox matter. (1851, 513)

Similarly, Simpson

could not doubt that the saturation of the bed-clothes, etc., with the discharges of a puerperal fever patient, might give the same disease to another puerperal patient who was laid in them. This, and one or two other circumstances, were enough to show that, for safety's sake, it was always well to act upon the possibility of the clothes even of the medical attendant being thus a medium of contagion. (1851, 512).

It is natural for us to interpret these passages as being about some sort of material contaminant, akin to bacteria, and so to us the evidence seems to suggest quite overwhelmingly that it was some specific matter on the attendants' hands or instruments that was responsible for disease transmission, and that this matter could survive on other materials, such as an attendant's gloves, just as we know pathogenic microorganisms do. We might even wonder, on this view, why there was such opposition to early germ theories and why it took so long for such views to become established; after all, from our perspective, the evidence seems clear.

The situation at the time, however, was not that straightforward. For one, while the evidence suggested that there was a relationship between accoucheurs and patients, it did not specifically suggest that it was a material agent on the attendants' hands that was responsible for the passing on of the disease. Instead, the transmissibility hypothesis was entirely compatible with the view that there was something else about the doctors and midwives—such as a toxic odor clinging to them or their clothes—that was the culprit. In fact, it seemed that there was at least some evidence that the disease was sometimes transmitted through air and not touch, as is shown by the following episode, recounted by Simpson: "In some observation on the subject of the contagion of puerperal fever, Dr. Merriman states, that he once attended the dissection of a puerperal patient, but did not touch the body or any of the parts. The same evening he attended a lady in labour, and she was attacked with the disease" (1851, 512). In the same vein, there are many and extended discussions about various fumigation practices for clothing, whitewashing sickrooms, and so on. And while to us the many references to using chloride of lime might suggest something like disinfection or the destruction of microbial matter, in the eighteenth and nineteenth centuries chloride of lime was a popular choice because it was known to work against smells (Parsons 1978). Thus, even the transmissibility hypothesis involved some of the basic principles of the miasma theory: it is, at least in some cases, a type of noxious air that is causing puerperal fever, even if this noxious air is not always atmospheric in origin but can sometimes be generated by the disease victims themselves.

Thus, a belief in disease communicability did not entail a corresponding rejection of the miasma theory, and it is partially for this reason that it is a mistake to identify such beliefs with early germ views. The converse also holds: it is not warranted to regard the miasma theory as an essentially anti-contagionist view, since there were plenty of miasmatists who thought that diseases could be passed on, even if this was not always the case.[8] And, however complicated the general debate about disease theory may have been, at least with respect to puerperal fever, many writers were clear that they thought the evidence could not support anything more specific than the general association between accoucheurs and patients. It is for this reason that many writers were silent on the details of the specific modes of transmission of puerperal fever; instead, what they were interested in first and foremost was to establish *that* the disease was communicable, and that there was a cause-and-effect relationship between specific practitioners and victims of the disease. Holmes, for example, even went as far as to say that he thought this sort of speculation was futile, since the evidence itself could not decide the matter one way or another:

> I shall not enter into any dispute about the particular *mode* of infection, whether it be by the atmosphere the physician carries about him into the sick chamber, or by the direct application of the virus to the absorbing surfaces with which his hand comes in contact. Many facts and opinions are in favour of each of these modes of transmission. But it is obvious that in the majority of cases it must be impossible to decide by which of these channels the disease is conveyed, from the nature of the intercourse between the physician and the patient. (Holmes 1855, 28)

It is also worth stressing that even those who believed in the transmissibility of puerperal fever thought that this was only one way in which people could catch the disease. Most held that, besides being transmitted by birth attendants, puerperal fever could also be caused in some way or another by atmospheric conditions. Holmes, for example, writes that "[i]t is not pretended that the disease is always, or even, it may be, in the majority of cases, carried about by attendants; only that it is so carried in certain cases. That it may have local and epidemic causes, as well as that depending on personal transmission, is not disputed" (1855, 20). Indeed, some sort of atmospheric explanation was thought to be necessary for explaining epidemics, which often happened in different parts of the country at the same time, or even in different countries. Thus, Simpson

> believed that we ought not altogether to forget the possibility of epidemic influences acting directly or indirectly in the causation of it. During the present century the disease had nearly, in two or three instances, as in 1819–20 and 1829, prevailed in most of the cities and lying-in hospitals of Europe. And it was difficult or impossible to account for this simultaneous existence everywhere, without believing that everywhere there was some general epidemic cause tending to its production (1851, 514).

The notion that some more general epidemic influence was at least sometimes at play could also explain the fact that puerperal fever was seasonal, with much higher morbidity and mortality rates during the winter than the summer months. This would have been difficult to account for on a view in which the only cause of puerperal fever was a previous case of the disease. Similar reasoning also seemed necessary in order to account for the sporadic version of puerperal fever, which always existed, although it had a much less extreme fatality rate: "Dr. Simpson observed, that no doubt sporadic cases of puerperal fever frequently did occur traceable to no contagion or any other cause capable of being averted" (Simpson, 1851, 515). More generally, given what was known about puerperal fever (and other diseases) at the time, multicausal views of disease causation had much greater explanatory power than an infection-only view, which left many questions unanswered, especially those about seasonality and concurrent outbreaks in different regions. And for many people, adding one more cause to an already long list of possible causes was not a big step, especially if that further cause involved the transmission of odors.

Upshot of Case Study

The case study shows that there are at least some instances in which it is plausible to think that the virtues make an epistemic contribution to our theories and hypotheses. Moreover, instead of abstractly suggesting the possibility of the virtues being epistemic in character, it offers a concrete scenario in which we can see specific virtues doing specific epistemic work: we see how some such virtues were in fact invoked in a case of real-life theory choice, and that and how they were used in order to argue for the *truth* of the transmissibility hypothesis—a distinctly epistemic context. Of course, the mere fact that they were invoked in such a context does not make them generally epistemic, even if the case study makes plausible the view that they might be. To establish more generally that the virtues make epistemic contributions to our theories, and to be clearer about the kinds of contributions they make, we need to study more examples, and to see how the virtues function in different scenarios. It might be, for example, that certain virtues are consistently given more weight than others or that under different circumstances people tend to invoke different virtues. However, this is something we can shed light on only by trying to understand how the virtues were used by actual people in actual debates, not by merely thinking about them in the abstract. Even if, for example, we could somehow determine, perhaps through some kind of a *priori argument*, that virtue 1 ought to carry more epistemic weight than virtue 2, this cannot illuminate how the virtues function in instances of scientific justification unless it is also the case that actual scientists' reasoning mirrors this hierarchy. After all, science has been successful the way it actually did happen, not the way it ought to have happened.

Moreover, the case suggests not just that it is important to study how the virtues were used in actual scientific debates, but also that we ought to study how they were used to resolve those debates. This, too, represents a shift in the debate about virtues, since usually when cases are invoked at all in this debate, what we see are arguments seeking to link a theory's having particular virtues with that theory's truth(likeness). Since realists typically want to argue for the view that the virtues are truth-conducive—in other words, that theories having the virtues are more likely to be true than theories lacking them—this makes sense. As a result, the typical strategy for the realist is to look at theories that we now take to be (approximately) true, find theoretical virtues in those theories, and then show that these virtues do not appear in the corresponding competitors. That is to say, the strategy is to show that "winning" theories have virtues, whereas "losing" theories seemingly do not. The usual way to show this is to point to a particular virtue and to show that one theory or hypothesis exhibits that virtue, whereas its rival does not—rather like in the case of puerperal fever, where we might point to the fact that the transmissibility hypothesis could explain why the disease followed certain practitioners and the non-transmissibility view failed to account for this connection. So this seems like an instance of a case where the "winner" has a virtue and the "loser" does not. And, on the face of it, this would appear to be one piece of evidence in favor of the realist view, since we now know that the transmissibility hypothesis turned out to be true and its rival did not. Thus it seems that what the realist needs to do to gain further support is simply to find more such cases.

There is, however, a problem with this strategy, and it is this: showing that particular virtues were not instantiated by the losing theory in a particular context does not show that the theory did not have any virtues. Just because a theory cannot, say, explain a certain phenomenon and another theory can, does not mean that there might not be other phenomena on which the first theory does better. In the case of puerperal fever, for example, we saw that the miasmatic view could explain certain states of affairs that a pure contagionist view could not, such as well-known facts about disease seasonality and concurrent epidemics in different locations. So even if we can show that certain theories do well on certain virtues with respect to certain phenomena, this does not mean that theories that don't do well on those virtues with respect to those phenomena don't do well with respect to others. The very same theories that fail in one circumstance might instantiate other virtues—or even the very same virtues—in another.

In the debate about puerperal fever, that seems to have been exactly the situation. It is easy to look back at the debate about transmissibility and find something like a debate between contagionists and anti-contagionists, with the contagionist hypothesis having virtues, winning out, and looking like a precursor to the germ theory, and with the losing anti-contagionist view lacking those virtues. However, a closer look reveals that this is not even the right opposition: while it is true that the transmissibility hypothesis had many virtues, it only claimed that puerperal fever was sometimes communicable. It remained silent on the details of the different possible modes of transmission, with most of its proponents believing that puerperal fever, when transmitted, was at least sometimes transmitted through noxious odors. Further, most of its proponents were also adherents of the larger miasma theory, and so communicability through smells was added as merely one of many on a long list of causes that still included as central causal agents atmospheric and telluric conditions. What's more, the same virtues that were invoked in arguing for the (true) transmissibility hypothesis were also invoked in arguing for the (false) miasma theory more generally. Moreover, they were often invoked by the very same people, and so the difference here cannot be accounted for by generally better and generally worse groups of scientists.

However, the fact that people on all sides of the debate appealed to the virtues, and did so in contexts that were distinctly epistemic, should tell us something. First, it means that people with different views agreed, at least in principle, on some of the standards by which they ought to judge their hypotheses, even if they disagreed about which theories or hypotheses best exemplified these standards. Second, it means that the question, at least in this case, was not so much about whether a certain theory did or did not have virtues as about how to adjudicate among a number of competitors all of which had (some of) the same virtues. What we ought to look at is how these often complicated and sophisticated debates, virtues and all, got resolved. How were different virtues used by whom to argue what? Often such debates took decades to settle, as was the case with the debate between the miasma and germ theories. How did this decade-long transition occur? How was it that people who held at least some of the same epistemic standards changed their minds? In response to what arguments did their views evolve? How exactly did people adjudicate between different hypotheses, all of which were supported by some virtues? Were there virtues that were systematically held to be more important than others? Did certain virtues always dominate others? Were there cases of theories that had different virtues, without overlap? And so on.

Note that one consequence of this is that the virtues being epistemic and their being truth-conducive do not necessarily go together. It might be true that theories that have virtues are more likely to be true than theories that don't have any, but the case of puerperal fever gives us at least some reason to doubt that actual, live scientific scenarios involve this kind of choice. Instead, in such scenarios, the debate is between theories both of which have at least some virtues. The case of puerperal fever is, of course, only one case, and while here the hypotheses on all sides exhibited virtues, this is no guarantee that other cases are similar. Other debates in nineteenth-century medicine appear analogous, but even they could of course be unrepresentative of science at large. However, reading historically informed and detailed discussions of scientific debates and controversies gives the impression that it is not unreasonable to suppose that this state of affairs is common. But, however, this may be, the only way to find out is by studying more virtues in more detail in more episodes in the history of science.

I want to end by noting that, even on a realist view, it would be surprising if (mature) discarded theories ended up not having any virtues whatsoever. After all, the history of science is a progression of theories that get better with time, even if only according to a vague, imprecise, and not very well-defined notion of "better." But this means that theories that turned out to be losers were at some point winners and so, even (or perhaps especially) on a realist view, we should expect such theories to exhibit some virtues and their proponents to have invoked good arguments in their favor. Thus, contra the realist, the truth-conducive and epistemic natures of the virtues can come apart: being epistemic does not make a virtue truth-conducive.[9] But it also means, contra the anti-realist, that the virtues are not merely pragmatic and that debates presupposing this dichotomy do not mirror the historical realities of theory choice, whatever exactly they may be.

Conclusion

The case of the debate about the transmissibility of puerperal fever in the mid-nineteenth century shows that no good answer to the virtue question can be had without taking into account detailed empirical studies of the virtues in action. In order to ascertain whether there are virtues that contribute to the epistemic standing of our scientific theories, we need to examine cases in which we can observe the virtues at work. What the case study in this chapter suggests is that the virtues are at least sometimes put to epistemic use. However, being epistemic in character is no guarantee of a virtue's truth-conduciveness, as we have seen. It is for this reason that the virtue question, as it appears in the realism debate, can benefit from having its focus shifted: instead of merely examining whether or not certain theories and hypotheses have specific virtues, we ought to put more effort into determining how disputes among theories both of which have virtues play themselves out. As a result, thinking about epistemic virtues—and other epistemic properties that scientific theories might possess—goes much beyond resolving issues to do with the realism debate. Rather, it is the beginning of an account that seeks to shed light on more general questions about the epistemic status of scientific theories, such as questions about what sorts of factors make epistemic contributions to our scientific hypotheses and in what ways they do so.

Notes

* Many thanks to Tom Stapleford and Emanuele Ratti for their invitation to contribute to the workshop "Science, Technology, and the Good Life: Perspectives on Virtue in Science and Technology Studies," University of Notre Dame, April 2018, and to

this volume. This chapter is adapted, by permission from Springer, from "Epistemic Virtues and the Success of Science," in *Virtue Epistemology Naturalized*, edited by Abrol Fairweather, 247–68 (Cham: Springer, 2014).

1. For more details on this argument, see Psillos 1999 and Tulodziecki 2012 and 2013a. Since my main focus here is not underdetermination, I've opted to deal with it only briefly, so I can devote more time to a discussion of the virtues. For more on the scientific realism debate, see Saatsi 2017.

2. Although see Schindler 2018.

3. Simplicity (and, relatedly, parsimony) are particularly notable in this respect. See, for example, Sober 1988, 1996, 2002a, 2002b; Forster 1995a, 1995b; Forster and Sober 1994; Kelly 2007a, 2007b.

4. A noteworthy exception is McAllister (1989, 1996), who draws a distinction between virtues as indicators of truth and virtues as indicators of beauty, the former being formulated *a priori*, while "aesthetic criteria are inductive constructs which lag behind the progression of theories in truth-likeness" (1989, 25).

5. There are widely diverging accounts as to what percentage of women actually contracted the disease. For more details on the disease and its history, see Loudon 1986, 1992, 2000.

6. See Hamlin 1992 for a discussion of the distinction between predisposing and exciting causes.

7. This choice of virtues is motivated by the fact that these three are virtues on which there seems to be agreement among most realists (as opposed to simplicity or elegance, say), but plenty of others can be found.

8. The dominant view was so-called contingent contamination; for details, see Hamlin 2009 and Pelling 1978.

9. Talking about conduciveness to approximate truth does not help here, since at a point in time when scientists are engaged in a controversy, we still don't know how to single out one theory as being more likely to be approximately true than another, if both have virtues. Of course, this doesn't make the realist position untenable; it just means that realists ought to focus on how to adjudicate between the virtues in such cases, as I have suggested.

References

Baldwin, P. 1999. *Contagion and the State in Europe, 1830–1930*. Cambridge: Cambridge University Press.

Bird, A. 2007. "Inference to the Only Explanation." *Philosophy and Phenomenological Research* 74, no. 2: 424–32.

Bird, A. 2010. "Eliminative Abduction: Examples from Medicine." *Studies in History and Philosophy of Science* 41: 345–52.

Feest, U., and T. Sturm, eds. 2011. "What (Good) Is Historical Epistemology?" *Erkenntnis* 75, no. 3: 285–302.

Forster, M. R. 1995a. "The Golfer's Dilemma: A Reply to Kukla on Curve-Fitting." *British Journal for the Philosophy of Science* 46, no. 3: 348–60.

Forster, M. R. 1995b. "Bayes and Bust: Simplicity as a Problem for a Probabilist's Approach to Confirmation." *British Journal for the Philosophy of Science* 46, no. 3: 399–424.

Forster, M. R., and E. Sober. 1994. "How to Tell When Simpler, More Unified, or Less Ad Hoc Theories Will Provide More Accurate Predictions." *British Journal for the Philosophy of Science* 45, no. 1: 1–35.

Gillies, D. 2005. "Hempelian and Kuhnian Approaches in the Philosophy of Medicine: The Semmelweis Case." *Studies in History and Philosophy of Biological and Biomedical Sciences* 36, no. 1: 159–81.

Gordon, A. (1795) 1849. "A Treatise on the Epidemic Puerperal Fever, Etc." In *Essays on the Puerperal Fever and Other Diseases Peculiar to Women*, edited by F. Churchill. London: Sydenham Society.

Hamlin, C. 1992. "Predisposing Causes and Public Health in Early Nineteenth-Century Medical Thought." *Social History of Medicine* 5, no. 1: 43–70.

Hamlin, C. 2009. *Cholera: The Biography*. New York: Oxford University Press.

Hempel, C. G. 1966. *Philosophy of Natural Science*. Englewood Cliffs, NJ: Prentice-Hall.

Holmes, O.W. 1855. *Puerperal Fever as Private Pestilence*. Boston: Ticknor and Fields.

Holmes, O. W. 1892. *The Works of Oliver Wendell Holmes: Medical Essays 1842–1882*, vol. 9. Boston: Houghton Mifflin.

Kelly, K. 2007a. "A New Solution to the Puzzle of Simplicity." *Philosophy of Science* 74: 561–73.

Kelly, K. 2007b. "How Simplicity Helps You Find the Truth Without Pointing at It." In *Induction, Algorithmic Learning Theory, and Philosophy*, edited by M. Friend, N. Goethe, and V. Harizanov, 111–43. Logic, Epistemology, and the Unity of Science, vol. 9. Dordrecht: Springer.

Kukla, A. 1993. "Laudan, Leplin, Empirical Equivalence, and Underdetermination." *Analysis* 53: 1–17.

Kukla, A. 1996. "Does Every Theory Have Empirically Equivalent Rivals?" *Erkenntnis* 44, no. 2: 137–66.

Kukla, A. 1998. *Studies in Scientific Realism*. New York: Oxford University Press.

Laudan, L., and J. Leplin. 1991. "Empirical Equivalence and Underdetermination." *Journal of Philosophy* 88: 449–72.

Lipton, P. (1991) 2004. *Inference to the Best Explanation*. London: Routledge.

Loudon, I. 1986. *Medical Care and the General Practitioner: 1750–1850*. Oxford: Oxford University Press.

Loudon, I. 1992. *Death in Childbirth*. Oxford: Clarendon.

Loudon, I. 2000. *The Tragedy of Childbed Fever*. Oxford: Oxford University Press.

McAllister, J. W. 1989. "Truth and Beauty in Scientific Reason." *Synthese* 78, no. 1: 25–51.

McAllister, J. W. 1996. *Beauty and Revolution in Science*. Ithaca, NY: Cornell University Press.

Parsons, G. 1978. "The British Medical Profession and Contagion Theory: Puerperal Fever as a Case Study, 1830–1860." *Medical History* 22, no. 2: 138–50.

Pelling, M. 1978. *Cholera, Fever and English Medicine, 1825–1865*. New York: Oxford University Press.

Psillos, S. 1999. *Scientific Realism: How Science Tracks Truth*. London: Routledge.

Psillos, S. 2004. "Tracking the Real: Through Thick and Thin." *British Journal for the Philosophy of Science* 55, no. 3: 393–409.

Saatsi, J., ed. 2017. *The Routledge Handbook of Scientific Realism*. New York: Routledge.

Schindler, S. 2018. *Theoretical Virtues in Science: Uncovering Reality Through Theory.* Cambridge: Cambridge University Press.

Scholl, R. 2013. "Causal Inference, Mechanisms, and the Semmelweis Case." *Studies in History and Philosophy of Science* 44: 66–76.

Simpson, J. Y. (1851) 1871. *Selected Obstetrical & Gynaecological Works of Sir James Y. Simpson: Containing the Substance of His Lectures on Midwifery.* Edited by J. Watt Black. Edinburgh: Adam and Charles Black.

Sober, E. 1988. *Reconstructing the Past: Parsimony, Evolution, and Inference.* Cambridge, MA: MIT Press.

Sober, E. 1996. "Parsimony and Predictive Equivalence." *Erkenntnis* 44, no. 2: 167–97.

Sober, E. 2002a. "Instrumentalism, Parsimony, and the Akaike Framework." *Proceedings of the Philosophy of Science Association* 69, supp. 3: S112–23.

Sober, E. 2002b. "What Is the Problem of Simplicity?" In *Simplicity, Inference, and Modelling,* edited by A. Zellner, H. Keuzenkamp, and M. McAleer, 13–32. Cambridge: Cambridge University Press.

Tulodziecki, D. 2012. "Epistemic Equivalence and Epistemic Incapacitation." *British Journal for the Philosophy of Science* 63, no. 2: 313–28.

Tulodziecki, D. 2013a. "Underdetermination, Methodological Practices, and Realism." *Synthese* 190, no. 17: 3731–750.

Tulodziecki, D. 2013b. "Shattering the Myth of Semmelweis." *Philosophy of Science* 80, no. 5: 1065–75.

van Fraassen, B. 1980. *The Scientific Image.* Oxford: Clarendon.

Whewell, W. 1858. *Novum Organon Renovatum.* London: John W. Parker.

PART IV

VIRTUES AND RESEARCH ETHICS

12

Integrating Virtue Ethics into Responsible-Conduct-of-Research Programs: Challenges and Opportunities

Jiin-Yu Chen

In 1989, prompted by highly visible revelations of research misconduct in the sciences, the National Institutes of Health (NIH), in conjunction with the Alcohol, Drug Abuse, and Mental Health Administration (ADAMHA), issued the first federal requirement that certain grant recipients develop a plan for instruction in the responsible conduct of research (RCR) (NIH 1989). By 1997, the National Science Foundation (NSF) introduced its first RCR education requirement in the Integrative Graduate Education and Research Traineeship program (Steneck and Bulger 2007). Since then, both agencies have expanded the scope of their RCR education requirements. The NSF now requires all trainees it supports to receive RCR education, including undergraduates, graduate students, and postdoctoral fellows (NSF 2009). The NIH RCR requirements now also include many levels of trainees, as well as more specific guidelines on content, duration, and format of RCR education, although there is still room for interpretation (NIH 2009). Additionally, the United States Department of Agriculture (USDA) now has an RCR education requirement (USDA 2018). For all of these agencies, the responsibility for developing and implementing these RCR education programs, as well as ensuring compliance, falls to the institution.

This chapter will examine some of the RCR education programs that institutions have developed to meet the federal requirements, as well as the obstacles that RCR education programs typically face. Many programs seek to cultivate the development of good ethical judgment as a way of fostering integrity in research. For this goal, virtue ethics can provide a helpful framework by focusing on the cultivation of virtues in the moral agent's character for a life well lived. Virtue ethics can be brought to bear on research by highlighting the virtues researchers must cultivate to produce good science. Developing RCR education programs that incorporate virtue ethics draws attention to the importance of this cultivation and connects integrity in research with the researcher's character.

Jiin-Yu Chen, *Integrating Virtue Ethics into Responsible-Conduct-of-Research Programs: Challenges and Opportunities*
In: *Science, Technology, and Virtues*. Edited by: Emanuele Ratti and Thomas A. Stapleford, Oxford University Press.
© Oxford University Press 2021. DOI: 10.1093/oso/9780190081713.003.0013

Common Obstacles Faced by RCR Education Programs

Undoubtedly, the federal requirements mandating RCR education have resulted in the widespread proliferation of RCR education programs. In many ways, the impetus behind these requirements is commendable in its attempt to systematically familiarize generations of researchers with common ethical issues related to their training and work and to encourage early deliberation on these issues. The hope is that early familiarity with some of the issues may facilitate resolution of ethical conflicts sooner rather than later, as conflicts that are left unaddressed until later often present more challenges. Another benefit is that those affected would know where to go to seek help.

Institutions have flexibility in developing many features of their RCR education programs, but this flexibility has also been a double-edged sword, and a multiplicity of different approaches to RCR education programs has emerged (Mastroianni and Kahn 1998, 1999; Heitman and Bulger 2005; DuBois et al. 2010). While the content covered by RCR education programs tends to be relatively similar, typically covering areas such as research misconduct, authorship, publication ethics, the mentor-mentee relationship, data management, and peer review, these programs vary in a number of other respects. Institutions may meet the minimum federal compliance requirements by requiring only those students and postdocs receiving NIH, NSF, or USDA funding to complete their RCR education program. Alternatively, they may establish their own institutional policies in addition to the federal requirements, increasing the number of people who must participate in and complete their RCR education program. Further, institutions may offer only one option for completing the RCR requirement, which could be administered centrally or based within departments, or offer multiple paths for meeting the requirement. The format may be online, in-person, or a mixture, and the in-person component can vary in terms of who organizes and teaches those sessions (Staff? Faculty? Both?), the duration of each session (One hour? Six hours?), who can access those sessions (Only doctoral students? Those from specific disciplines? Everybody?), and how frequently they are offered (Many times in the academic year? Once during orientation?). Because the RCR requirements apply to many different kinds of institutions with varying research profiles, focuses, and populations, flexibility has been important in allowing institutions to develop RCR education programs tailored to their particular situations. An RCR education program that addresses the ethics concerns of a large academic medical institution would not likely be appropriate for a small liberal arts university whose research is primarily in the social sciences. Additionally, some RCR education programs may not be feasible for some institutions, depending on resource and available support.

As an example, the University of Pittsburgh has developed the RCR Center, which holds frequent, discussion-based, hour-long workshops on various RCR topics. These utilize expert speakers and are broken into core instructional topics and specialized topics. The workshops are attended by graduate and undergraduate students, postdoctoral fellows, junior faculty, medical students, and research staff and coordinators, many of whom are not seeking federally mandated RCR education (Schmidt et al. 2014). In contrast, Georgia Tech has broadened the scope of its RCR requirements, mandating that not only those students and postdocs who receive funds from the NSF, NIH, or USDA complete its RCR program, but also that all doctoral students and master's thesis students, regardless of funding source, complete the program (Georgia Tech n.d.). This requirement is fulfilled through a combination of online training and in-person educational sessions dependent upon level and funding source. Some of the in-person sessions are open to all who need to complete the requirement, while others are offered through certain departments and are open only to their students. Similarly, Michigan State University requires all of its students at both the master's and doctoral levels to fulfill its RCR requirements (Michigan State University n.d.). The requirement is met through a combination of online modules and in-person discussions, with the doctoral students required to participate in annual refresher training after their third year. The in-person sessions, a series of workshops focused on specific topics, are offered throughout the academic year.

Other approaches to RCR education focus less on specific topics and more on the decision-making process and the many factors to consider. Mumford et al. (2008) and Kligyte et al. (2008) developed a sensemaking approach in which, over the course of two days, they presented and discussed modules that established a framework for ethical decision-making. These modules included learning about guidelines, complexities in ethical decision-making, personal biases, reasoning errors, and field differences. While these studies reported the results of an early version of their program, the University of Oklahoma (OU) developed it into the Professional Ethics Training–Responsible Conduct of Research Program, and expanded it to all OU graduate students with an assistantship appointment who seek to qualify for a tuition waiver (OU Graduate College n.d.).

Differences in what is federally required for RCR education programs add to the lack of clarity. The preceding examples demonstrate considerable variation in how institutions have developed their policies and procedures to meet the RCR education requirements. If an institution adheres to what is minimally required by each federal agency, these discrepancies can lead to inconsistencies in the RCR education that trainees at the same institution receive. For example, the NIH requires trainees to receive at least eight hours of in-person RCR education

(NIH 2009); however, neither the NSF nor the USDA specifies length or format (NSF 2009; USDA 2018).

In examining how RCR education plans developed in response to the NSF requirements, Phillips et al. (2018) analyzed publicly available NSF RCR training plans at US institutions with "very high research activity" in the Carnegie Classification. They found that many institutions offered only a single path for meeting the RCR education requirement, which was frequently delivered through online training. This online-only format is not sufficient to meet the NIH requirements. While it does meet the NSF requirements, sole reliance on an on-line format is not recommended as a best practice for RCR education programs, which should include active participation (CGS 2008; NASEM 2017). When institutions did offer multiple paths for meeting the requirement, the study found significant differences between what was offered and what was required in order to meet the federal mandates (Phillips et al. 2018). That is, institutions offered several ways for trainees to receive RCR education, typically varying in format (in-person versus online), which could result in very different RCR education experiences among trainees. Notably, the majority of the institutions surveyed also receive NIH funding and have RCR education programs in place to meet the NIH's RCR requirements. While there is some evidence that institutions are developing RCR education programs that meet the minimum requirements of each agency, a number of other institutions have developed institutional RCR requirements that go beyond the federal requirements, such as including trainees who do not receive federal funds. Resnik and Dinse (2012) found that slightly fewer than half of the top-funded research institutions in their survey required only those who are federally mandated to complete their institution's RCR education program, while the rest required other individuals to complete RCR training as well, such as all students who participate in externally funded research or all doctoral students. They also found that institutions with medical schools were more likely to have institutional requirements beyond the federal mandates.

A result of this variety is a lack of agreed-upon rationales for and goals of RCR education programs, which have "come to include a wide-ranging mix of knowing and following the rules, being a moral person, having good character, exhibiting good ethical judgment, and acting with integrity and responsibility" (Kalichman 2007, 870). For the long term, Kalichman argues that RCR education programs should seek to reduce incidences of research misconduct and promote a shared vision of practicing science according to accepted standards and norms (Kalichman 2007). Regardless of the objectives of RCR education programs, how well the curricula meet these objectives is uncertain. While explicit educa-tion on research misconduct and the standards and norms of scientific practice is important, particularly to those unfamiliar with those concepts, a number of factors affect how closely one adheres to those norms, such as those related to the individual's character and environment (Kalichman 2007). Additionally, RCR education programs are only one small component of a researcher's training and career. The time spent in and content of these programs can easily be lost in the daily shuffle of one's work.

Many RCR education programs face a lack of resources and institutional support. There are a number of publicly available resources for teaching RCR, but institutions may lack faculty or staff who are motivated and well-prepared to teach it (Kalichman 2014).[1] Furthermore, inexperienced or unmotivated RCR instructors may not be aware of the available material or best practices regarding instructional tools or format for RCR education. And while RCR education programs do not usually require large amounts of funding, some amount of reliable funding is needed for sustaining RCR education programs in the long term. Increased funding could allow RCR instructors to become better trained, to cover associated administrative expenses such as room fees and printing costs, or to pay for experienced speakers to present on an RCR topic, all of which would signal institutional support for RCR education programs. Kalichman proposes that 0.1 percent of research direct costs be used for the development and maintenance of RCR education programs (Kalichman 2014).

Further, it is sometimes unclear who holds responsibility for RCR education at an institution—is it the RCR instructor, the administration, faculty, or someone else? Arguably, the responsibility of familiarizing researchers with the norms and standards of the practice lies with all parties, but institutions should be cognizant of how RCR education programs may inadvertently partition discussion of and reflection upon research ethics from the rest of the research enterprise. As previously discussed, some institutions require only those trainees who are receiving NIH, NSF, or USDA funding to complete their RCR education program. This sends the message that education in ethical and responsible research practice is merely a hurdle to be cleared, and that RCR education programs are something that institutions must provide because of the federal requirements, not because they communicate valuable knowledge for all trainees.

While the problems that commonly plague RCR education programs certainly present obstacles, they may be addressed through a program's structure and position within an institution. I describe the program that I developed and administered Boston College, to give an in-depth example of how an RCR education program might begin to address some of the challenges. For RCR education, Boston College established the Research and Scholarship Integrity (RSI) program. Aimed at researchers and scholars still in training, the RSI program seeks to familiarize these individuals with the ethical issues they may encounter over the course of their careers and to create spaces for reflection upon and discussion of these potential situations and how they might be addressed. The RSI program covers topics most frequently included in RCR education programs, such as research misconduct, responsibilities toward society, authorship and publication ethics, and conflict of interest, and includes a few additional topics such as copyright, frameworks for addressing ethical issues, and race and gender bias in academia. These subjects are covered through a series of in-person sessions, with no online component.[2] Although resource-intensive, this in-person structure enables large- and small-group deliberations about each topic. Some of these sessions are required for all participants, whose topics focus on areas that are broadly applicable across disciplines. Other sessions are more targeted to particular disciplines and areas, and participants must attend a set number of sessions, selecting from the annual offerings. As is the case with some of the other institutions mentioned, the RSI program is required for all entering doctoral students and new postdoctoral research fellows, regardless of discipline or funding source. Completion of the requirements of the RSI program is recorded in a graduate student's transcript, while postdocs are given a certificate of completion.

In separating the RSI program from federal RCR mandates while still meeting them, the university's goal is to send the message that works in all sectors of academia entails unfamiliar ethical terrain and that understanding and grappling with the ethical issues that may arise is a critical component in one's development as a researcher and scholar. The use of solely in-person seminars creates space for rich discussion of each session's topic among the attendees and presenters. Administration and management of the RSI program are centrally located in the Office of the Vice Provost for Research, as opposed to other models that distribute responsibility and oversight for RCR education across several departments and offices. With this administrative support, the RSI program receives a budget that covers various logistical considerations such as room fees, the cost of printing informational brochures, and food for participants.

Tensions Stemming from Mandatory RCR Education Programs

In many ways, the RSI program at Boston College avoids some of the challenges that many RCR education programs confront. It receives many of the resources that enable a broadly required, in-person RCR education program. The flexibility built into the structure is attentive to the variety of ethical issues researchers in different areas face and the competing demands on the participants' schedules. However, despite the support, the RSI program faces some fundamental challenges and obstacles that plague other RCR education programs as well. Regardless of the resources available to a program, the expertise of the

instructors, or an institution's support, the compliance aspect presents fundamental challenges to accomplishing program goals. Compliance is not optional;
institutions that accept NIH, NSF, or USDA funds must develop RCR education programs. However, compliance can also connote a sense of hindrances
without benefit, of meaningless boxes to be checked, of byzantine and vaguely
threatening bureaucracies. And, as previously discussed, some institutions have
responded to the RCR requirements by offering what is minimally acceptable for
particular funders. However, RCR education programs are shaped by a number
of factors, and we must be careful in attributing reasons for their current structure and shape.

Regardless of any institution's actual RCR education program and the
conceptions that surround all forms of compliance, mandatory ethics programs
may suffer from other negative associations. They may be seen as accusatory and
suspicious. In part, the federal regulations were implemented to prevent research
misconduct, but implementation has resulted in the perception that researchers
do not know how to behave ethically and need outside guidance—as if their
moral compasses need correcting.[3] Understandably, researchers take offense
at this interpretation. Further, mandatory ethics programs may also be seen as
rules-oriented, instead of as promoting deep and thoughtful exploration into the
various contexts and factors at play in ethical issues, the potential responses to
and resources for navigating those issues, and ways of preventing issues from
arising. Related to the perception that researchers don't know how to behave ethically is the fact that ethics programs are sometimes seen as a litany of self-evident
rules such as "don't cheat" and "do keep accurate lab notebooks and records" instead of facilitating discussions about what counts as cheating and what belongs
in lab notebooks. These conceptions combine to give the impression that RCR
education programs operate from thin conceptions of ethics instead of delving
into thick ethical discourse.

In addition to the negative connotations associated with RCR education
programs, many questions arise about the effectiveness of these programs.
Plemmons and colleagues (2006) found that RCR courses increased participants'
knowledge but did little to change their skills in addressing the ethical dimensions
of their research or their attitudes toward responsible research. Schmaling and
Blume (2009) surveyed graduate students at the beginning and end of a semester-
long RCR course and found that knowledge about RCR increased but moral
judgment did not. Antes et al. (2010) found that RCR instruction improved
participants' metacognitive reasoning strategies in some areas, such as awareness
of the situation and consideration of personal motivations, but they also found a
decline in other areas such as seeking help, anticipating consequences, and considering others. These results are mixed, indicating some improvements at best,
but also possible harms at worst. Of course, comparing RCR education programs

against each other is difficult given the wide variety. And the success of any program should be relative to its goals, which is also difficult given the lack of clarity surrounding those goals, as discussed earlier. Taken together, these studies indicate that the overall effectiveness of these programs is questionable.

RCR education programs then face a problem. On a large scale, they largely arose from a compliance-driven, regulatory background that continues to shape their direction and attitudes toward them. However, the goals of many RCR education programs are aspirational, seeking to promote research integrity through changing behavior, character, and judgment. They do this by utilizing education to increase knowledge and reflection about the myriad ethical issues researchers and scholars may encounter. These aspirational goals are not opposed to the compliance component, but neither do the two parts fit together easily. How are we to respond to this tension? How might a thick conception of responsible research and research integrity be promoted while also maintaining compliance with federal regulations?

One response might be to implement additional regulations or to make the existing regulations more robust and specific. In some ways, such changes could be beneficial. They could promote harmony between the different federal regulations and give institutions and researchers more direction in what federal agencies mean by "responsible conduct of research." More robust regulations could indicate that the federal agencies are taking RCR seriously. However, increased regulation must strike a fine balance and not go too far by taking away the flexibility institutions have in developing RCR education programs that address the needs of their research and scholarly communities. Regardless of the potential benefits of increased regulation, the tension between compliance and integrity continues.

Examining Research Through Virtue Ethics

Given the myriad of challenges and questions that RCR education programs face, one might wonder if they are worth continuing. The answer is yes, they certainly are. The positive features may be easy to overlook because the baseline for thinking and talking about research ethics has moved, partly because of the RCR education programs themselves. Such programs currently exist on a large scale because their existence is federally mandated for many institutions; without such mandates, many institutions would not have them. Even though they were prompted by the federal mandates, institutions must dedicate some resources toward them and carve out space in their students' and postdocs' training. This creates a place in the curriculum for education in research ethics, signaling the importance of this work. Further, the topics RCR education programs typically cover are wide-ranging, encompassing a number of areas that can threaten the integrity of research and others that examine the research environment and the responsibilities researchers have toward society. This wide scope promotes a broad understanding of what research ethics entails and an awareness that ethical issues can arise in many different ways. Despite the challenges the federal mandates present, they establish minimums for acceptable RCR education programs, which is arguably better than having no minimum at all. It is now commonplace in some disciplines to discuss responsible conduct of research as part of one's development as a researcher. Thus RCR education programs have increased the visibility of research ethics issues and potentially researchers' familiarity with those issues and comfort level in addressing them.

RCR education programs have room for improvement, but what they have accomplished, and the space they create for fostering improvements in how research ethics and integrity are taught and discussed, should not be discounted. As noted, the approaches RCR education programs use to teach ethics are highly varied. Many likely draw lightly upon various moral frameworks by examining ethical issues, weighing the harms against the benefits, elucidating responsibilities and duties, and setting forth principles to abide by. However, these approaches overlook the role of the researcher's character in ethics and integrity and leave little space for the aspirational ideals of teaching research integrity. Virtue ethics focuses on some of the features that RCR education programs seek to develop, such as character and judgment, and brings in a perspective of striving toward an ideal. Prior work (Bezuidenhout 2017; Bezuidenhout and Warne 2018; Caruana 2006; Chen 2015; Chen 2016; Deming et al. 2007; Macfarlane 2009; Pellegrino 1992; Pennock and O'Rourke 2017; Pennock 2018; Resnik 2012) and other contributions to this volume have already detailed virtue ethics and discussed how it may frame the research enterprise, so I will only give a brief summary of virtue ethics and focus on how some of its main features can enhance RCR education programs.

Virtue ethics focuses on the careful and deliberate habituation of virtues. These virtues enable the moral agent to pursue her conception of *eudaimonia*, of flourishing. This pursuit of flourishing happens over the course of the moral agent's lifetime and is something she continually strives toward. This flourishing and the necessary cultivation of the attendant virtues are not static but are always developing, continually undergoing reexamination and refinement. A central feature of virtue ethics is the careful and deliberate cultivation of the moral agent's character. Not only does the moral agent identify herself as an honest person, but those familiar with her do as well, for she has consistently and reliably acted with honesty in a variety of different situations. And importantly, the moral agent's honesty stems from appropriate reasons, namely that the situation calls for honesty and the moral agent acts as such because it is consistent with her character and not from ulterior motives such as wanting to appear honest in order to curry favor with someone else.

In the research context, much of a novice researcher's training and development hinges on the successful integration of the virtues that are critical to good research. During their training new researchers are expected to come to understand how virtues such as honesty, curiosity, perseverance, and objectivity function within the research enterprise and how they are to exercise those virtues.[4] If we frame research as a type of "practice," as Alasdair MacIntyre (2007) defines the term, then the cultivation of these virtues is essential to the proper functioning of research. MacIntyre holds that practices have their own distinct and specific internal goods that define and shape their standards of excellence and future direction. An internal good of research is the production of knowledge about the natural world using accepted methods. Successful pursuit of internal goods and achieving the standards of excellence that define the production of good research are attainable only through cultivation of the necessary virtues. This cultivation entails practitioners' adoption of these virtues into their identities as researchers. That is, practices hinge on the practitioners' identifying with those virtues needed to sustain that practice. Researchers must come to view themselves as honest, curious, persistent, and objective people. In the course of their long training period, they not only learn how to uphold those virtues in their work but also are given the space to integrate the virtues into their character and hone them in ways that produce good research.

A central feature of virtue ethics is the moral agent's cultivation of *phronesis*—practical wisdom. While virtues such as courage, temperance, and justice are traits of character that are evidenced by the moral agent acting in those ways, *phronesis* is the virtue that enables the moral agent to correctly discern if a situation calls for courage, temperance, or justice. *Phronesis* also enables the moral agent to determine when and how that virtue should be exercised to the appropriate extent and degree. Thus *phronesis* is the capacity to assess a situation, examine it from different perspectives, gather up the relevant details, and decide upon a virtuous path. While *phronesis* is a virtue, it differs from courage, temperance, and justice in that the correct exercise of those virtues hinges upon the moral agent's cultivation of *phronesis*; it is the virtue that undergirds all the other virtues.

Phronesis enables researchers to produce good knowledge. The researcher exercises *phronesis* when she determines whether she should adjust her line of inquiry and, if so, how to interpret data correctly and to report her findings accurately. But *phronesis* also allows the researcher to recognize situations that may threaten her integrity, such as accepting funds from a company whose interests overlap with her work, and to pay attention to other areas important to the proper functioning of research, such as recognizing the needs of her students and postdocs and preparing them well for their next career stage. Deming et al. (2007) interviewed twenty-three senior researchers and research administrators about their methods of reasoning when faced with conflict. Many of them mentioned using features of practical wisdom, such as developing intuition through experience, and calibrating this intuition through reflection, skepticism, and open discussion with colleagues.

Integrating Virtue Ethics into RCR Education Programs

Virtue ethics focuses on cultivation of character and *phronesis*, which both have important roles to play in conducting one's research with integrity. RCR education programs are potential sites for fostering this cultivation. What are some ways they might incorporate virtue? How can they contribute toward researchers' development of good moral character and judgment? How can they foster researchers' broad cultivation of *phronesis*? These questions do not have easy resolutions, but some work has been done to incorporate virtue ethics into RCR education programs. Deming et al. (2007) suggest that RCR education programs should be adapted to help researchers cultivate *phronesis* in their research practice by presenting explicit and clear articulations of professional responsibility, instead of relying primarily upon rules and principles for ethical conduct. They suggest examining and discussing how individuals develop their intuition through training, observation, experience, and interactions with others.

Programs have been developed with an explicit focus on virtue ethics and how that approach can enhance research ethics. Pennock and colleagues (2017) have developed ways to introduce students to exemplary scientists' characters and the necessary virtues. They put forth three approaches—theory-centered, exemplar-centered, and concept-centered—based in Pennock's scientific virtue theory, which seeks to demonstrate "what virtues stem from the epistemic values that underlie scientific practice and methods" (Pennock 2018, 168). They pilot-tested the concept-centered approach by organizing dialogue-based workshops around a particular scientific virtue. The focus of each module was a prompt designed to create discussion among the participants, such as "Objectivity implies absence of values by the investigator." The facilitated discussion sought to explicitly connect these scientific virtues with responsible conduct. For RCR education, discussing what virtues are needed by exemplary scientists draws the focus away from rules that dictate what one should and shouldn't do, which can sometimes be seen as externally imposed, and instead recognizes those virtues as central for good research. This shift in focus also gives novice researchers ideals toward which they should strive in order to conduct excellent research; it can guide them toward cultivating the kind of person they want to be (Pennock 2018).

Pennock and colleagues' approach incorporates virtue ethics education into a formal, structured RCR curriculum. They have placed the character of the researcher and the cultivation of the necessary virtues at the focal point and developed approaches to explicitly explore what that entails. While their work takes a strictly virtue ethics approach, to the exclusion of other approaches, virtue ethics can be incorporated into existing RCR education programs. For the in-person component, many programs utilize case studies to examine details of the topic at hand. For example, authorship can be a contentious area, where heated disagreement arises over who should be included as an author and in what order. Consider the following case study:

> Sophia began work on a research question as part of her graduate work. Unfortunately, she ran into a number of obstacles, and her advisor instructed her to discontinue work on the question, especially since it was tangential to her main research project. Sophia graduated and left the lab, but she remains at the institution as a postdoc. Her advisor believes Sophia's research question shows promise and encourages another graduate student, Andrew, to pursue it. Andrew, with input and some experimental assistance from Sophia, is able to develop a solution that resolves many of the main obstacles, and he collects enough data for a publication. For this solution, Andrew utilized the expertise of a colleague in another lab, who also supplied materials that were used in the solution. In drafting the manuscript, Andrew wrote much of it, with Sophia writing some parts of the results and discussion sections and reviewing and editing all sections. Both Andrew and Sophia claim first authorship on the publication. Who do you believe should be first author? Who else should be included as an author and in what order?

In working out how authorship questions may be resolved, participants might discuss what kinds of contributions to manuscripts merit authorship, the importance of authorship for their careers, and what additional information they would like to have. Discussion facilitators can provide structure by raising questions that examine the situation from different perspectives. Is it fairer to give Andrew or Sophia first authorship? What are the harms in denying someone authorship or giving someone undeserved authorship? Are other options available? What are some of the rules and best practices that guide authorship decisions? These questions highlight different features of ethical situations, and a similar approach can be utilized to bring in a virtue perspective.

Some examples of virtue-ethics-centered questions for this scenario could include: For all of the individuals involved, what virtues are at stake and which actions support them? In what ways could honesty affect who is included on the authorship list and the order? Which actions and reasons for those actions would enable the discussion participant to become the kind of researcher who regularly and reliably demonstrates those virtues? More broadly, discussion facilitators could guide conversation toward examining the role of one's character in conducting research with integrity. What responsibilities do researchers have? What virtues must they exercise to meet those responsibilities? Are the ways they meet these responsibilities reflective of their character? In what ways could they work toward improvement? How closely are those virtues aligned with their identities as researchers? Incorporating these kinds of questions into RCR education programs modifies existing structures and does not require the discussion facilitator to have deep expertise. And it would highlight an often overlooked area of research ethics, the researchers' character and virtues, by placing these alongside other important considerations in ethical decision-making.

Importantly, both approaches discussed for integrating virtue into RCR education programs rely on dialogue-based workshops. This overlap highlights the need for discussion in ethical reasoning and the critical function that RCR education programs have in creating protected space and time that is explicitly dedicated to that discussion. RCR education programs can have important structural and organizational roles in the novice researcher's training for safe exploration and discussion of various ethical issues. In utilizing this space for a virtue-ethics approach, the ideal format is an in-person discussion. Through discussion, the participants engage with the norms of research, the nuances of those norms, and various perspectives on those norms. This discussion-based format is particularly important as a tool to cultivate the novice researcher's *phronesis*, providing additional perspectives and experiences beyond her own and deepening her understanding of virtues in science and how they function.

Relationships, Mentoring, and Virtue Ethics

While integrating virtue ethics into a formal RCR curriculum can be fruitful, relying on RCR education programs as the sole site for students' deliberation of research ethics is likely not enough to prompt the deep changes in character that virtue ethics espouses. Researchers' informal but still critical interactions with difficult ethical situations are crucial opportunities to foster the cultivation of researchers' character and the attendant virtues. These interactions occur not in isolation but through conversations with others—colleagues, peers, professors, and mentors. It is through these relationships that novice researchers develop and adjust how they pursue integrity in their research and further a practice's norms, values, and standards of excellence. Bezuidenhout (2017) employs a virtue-ethics perspective to examine the importance of relationships in science. This approach highlights features central to the research enterprise that are often overlooked, including responsibilities researchers have toward sustaining the scientific community through the research they conduct and their recognition of the emotional dimensions of their relationships with colleagues and students.

A person's cultivation of *phronesis* and character is facilitated through relationships, notably with role models and exemplars. When a novice enters a practice, she learns about its norms, values, and standards of excellence not only through explicit instruction but also through learning about the accomplishments of excellent practitioners and the virtues and character they hold. She also observes more experienced practitioners in action, noting how others react to their work as well as discovering exciting new directions for the practice and what is rewarded. Much of this knowledge is implicitly gathered, but it is critical to the researcher's success. This knowledge is also sometimes explicitly discussed, but usually between the novice and the experienced researcher in situations outside the formal curriculum. The more experienced researcher gives the novice insight, perspective, and advice, serving in a mentoring capacity.

The mentoring relationship is critical to novice researchers' development and success. Through this relationship, mentees develop the skills and habits necessary for excellent research in a broad sense, which includes conducting research ethically and responsibly. Bird (2001) further discusses the ethical dimensions of the mentor-mentee relationship, many of which are related to conveying professional standards for excellent research but are also inherently laden with power dimensions that must be carefully navigated. Through their actions and responses, mentors convey "the range of accepted practice, shared views regarding what is unacceptable, and how seriously members of the community regard deviations from what is considered acceptable" (Bird 2001, 462). However, these observed behaviors should be explicitly discussed and explained by the mentor, lest they be subject to misinterpretation. Some of the ethical dimensions of the mentor-mentee relationship include questions of access and fairness, stereotyping, and misuse of power.

Nakamura and Condren (2018) report on a survey of exemplar scientists who demonstrated a high level of ethical commitment. They found evidence that this commitment was transmitted to those the exemplars had trained: "mentees were shaped by working for years in a virtuous microcosm established by the mentor, who periodically interacted with individual mentees and modeled virtue day by day" (Nakamura and Condren 2018, 7). However, they also report on another survey of early-career scientists that asked whether any ethical lessons they received from their mentors played a role in their professional formation. That survey found a much lower level of connection. Instead, the early-career scientists focused primarily on how the skills and knowledge they needed to conduct the research had been transmitted from their mentors (Nakamura and Condren 2018). The reason for this discrepancy is unclear. One possibility may be that the exemplar scientists placed greater weight on the ethical issues related to the research and explicitly discussed them with their trainees.

A study by Titus and Ballou (2014) may provide further perspective on this discrepancy. They sought to better understand whether faculty members who self-identified as advisors or mentors implemented RCR in their interactions with their doctoral students and how they understood responsibility for RCR training. They found that mentors were more likely than advisors to report working with their doctoral students on critical RCR components. This difference in how mentors and advisors approach RCR not only indicates differences in conceptions of one's role and responsibilities toward students but also suggests that mentorship entails placing some importance on addressing RCR issues. However, this finding is complicated; Titus and Ballou also found that about half the faculty saw responsibility for RCR as lying with the institution instead of with faculty, indicating inconsistent understandings of RCR responsibilities. Teaching about the ethical issues that can arise in research is often seen as the institution's responsibility, but not necessarily as a responsibility of researchers to examine these issues with those they train. Taken together, the findings of these studies indicate that the environment in which one trains and learns about the practice is critical for the researcher's development of technical skills and also for responsible and virtuous practice. Both affirm that mentors play critical roles in this development.

Mentoring, then, is an important component in cultivating the necessary virtues for excellent research practice. How could this mentoring piece be effectively integrated into RCR education programs? Many programs, Boston College's included, address mentorship like any other RCR topic and offer a module or seminar on it, discussing various features of the mentoring relationship. However, other approaches focus on developing mentorship itself and using it as the foundation for discussing research ethics. Kalichman and Plemmons discuss the development of workshops for faculty on how they can incorporate discussions of research ethics into their interactions with those they train (Kalichman and Plemmons 2018; Plemmons and Kalichman 2018). After a multiphase process during which they consulted experts in research ethics and convened focus groups made up of graduate students and faculty, they developed a curriculum for and assessment of the workshop (Plemmons and Kalichman 2018). These workshops focused on providing attendees with the tools for holding conversations about research ethics, incorporating the use of codes, checklists, cases, individual development plans, and policies (Kalichman and Plemmons 2018). The authors received positive feedback from the workshop attendees and, upon surveying their trainees, found that many of the workshop attendees incorporated one of the approaches covered in their research environments after the workshop (Kalichman and Plemmons 2018).

Although the workshops were small and may have suffered from self-selection bias, such efforts show promise in incorporating good mentoring practices as effective mechanisms for RCR education. Many participants initially expressed discomfort with teaching and discussing RCR with their trainees, but reported being much more comfortable with doing that after the workshop. With this explicit focus on teaching mentors how to discuss RCR topics with their trainees, discussions on the cultivation of virtues necessary for good research could also be integrated into training sessions for faculty. The faculty are taught how they might discuss virtues in research and their importance in the responsible conduct of research with their trainees. This discussion not only reinforces the importance of responsible conduct to the trainees but also models to the trainees the role of mentors in RCR education.

While virtue ethics can be employed to reframe and deepen RCR education programs, it overlooks some areas and should not be used exclusively. An area of weakness for virtue ethics is that because it focuses on the moral agent's cultivation of character, it sets aside the structural and systemic factors within which the moral agent operates. While cultivating virtue is important regardless of setting, one's cultivation of virtue can be facilitated or hindered by factors in an institution's structure. Cultivation of the virtues necessary for the responsible conduct of research is communicated through implicit and explicit messaging on the importance of RCR education, by the allocation of funds, time, and space for those purposes, and by the modeling of best practices by mentors. Development of these virtues is hindered by a lack of resources and support for RCR education and also by an environment in which there is fierce competition for the scarce resources necessary to succeed in research. In examining student cheating and academic dishonesty, James Lang (2013) looks at some of the systemic pressures that contribute to students' decisions to cheat. He argues that while some students may be more likely to cheat than others because of personality features or life situations, there are also contextual factors in the classroom environment itself that contribute to cheating. He argues that environments with an emphasis on performance, high-stakes outcomes, extrinsic motivation for success, and a low expectation of success all place pressures upon students to cheat (Lang 2013). These features are also characteristic of the current research environment. And while the researcher makes decisions about conducting her research with integrity, systemic factors should not be overlooked. Further, institutions arguably have the responsibility to mitigate threats to integrity and to foster environments where the virtues can flourish. Because of virtue ethics' focus on the individual moral agent, its approach could be complemented by other approaches that emphasize the structures and environments within which one works.

Another obstacle a virtue ethics approach faces is that it does not resolve the tension between the aspirational aims of RCR education programs and the need for compliance. A virtue ethics approach can direct attention toward and help meet these aspirational aims, but it does not address the compliance component. With federal or institutional requirements for RCR education programs, compliance continues to be an issue. That is not to say the requirements don't serve an important role; they do. They set a baseline minimum for education in the responsible conduct of research for all who fall within the requirements. Without this baseline, who receives education and how would be much more variable— some would receive RCR education far above the minimum, while others would receive little or none. However, while virtue ethics can move the focus and content of RCR education programs away from the minimum, it does not resolve the issues of perception and negative connotations that compliance can bring.[5]

RCR education programs can have an important role in fostering research integrity by establishing some minimums for familiarizing researchers with ethical issues they may encounter and with the standards for conducting research well. For novice researchers, they can provide protected space and time for exploring the nuances of these issues. However, RCR education programs also face considerable difficulties, such as a lack of clarity and agreement about their goals, the variation in their forms and structures across institutions, the mandatory component required by federal agencies, and the connotations about compliance that such a mandate brings.

Despite these difficulties, many RCR education programs have aspirational goals for cultivating the researcher's character and judgment toward the pursuit of excellence in her field. Striving toward integrity is an aspirational endeavor that works well with a virtue ethics approach, which can frame research as an enterprise with certain virtues, habits, and skills that its practitioners must cultivate in order to reach excellence. This cultivation can occur not only through explicit teaching on the necessary virtues but also through modeling by respected mentors and observation by novice researchers. In addition to the typical teaching formats of RCR education programs, such programs could also incorporate spaces for the cultivation and exercise of good mentoring practices and use curricula to emphasize the connections between mentorship, excellence, and integrity in research. Virtue ethics, when brought together with RCR education programs, can shift the focus away from the mere fulfillment of mandatory requirements and demonstrate how the cultivation of character, virtues, and habits is critical for research practiced with integrity.

Notes

1. Some of the online resources include the Online Ethics Center for Engineering and Science (https://www.onlineethics.org), Resources for Research Ethics Education

(http://research-ethics.org), and the Committee on Publication Ethics (https://publicationethics.org).

2. Because of COVID-19, the RSI program was conducted virtually for the 2020–21 academic year. All sessions were synchronous, with live presentations and discussions, to replicate the in-person format as closely as possible.

3. This perception is not necessarily unwarranted, as many RCR topics and case studies focus on examples of bad behavior by researchers and scholars. However, that is the hazard of teaching ethics—that materials are the result of ethical transgressions—and it is through retrospective analysis that the factors that led to the transgressions are unpacked.

4. These are not the only virtues expected of researchers. Any list of virtues is subject to debate, and these are some of the virtues that many agree are important for good research, according to a *Nature* poll (Nature Careers 2016).

5. This issue of compliance being unresolved is not specific to virtue ethics. Other frameworks that could be applied to RCR education programs also face the same problem.

References

Antes, Alison L., Xiaoqian Wang, Michael D. Mumford, Ryan P. Brown, Shane Connelly, and Lynn D. Devenport. 2010. "Evaluating the Effects That Existing Instruction on Responsible Conduct of Research Has on Ethical Decision Making." *Academic Medicine* 85, no. 3: 519–26.

Bezuidenhout, Louise. 2017. "The Relational Responsibilities of Scientists: (Re)considering Science as a Practice." *Research Ethics* 13, no. 2: 65–83.

Bezuidenhout, Louise, and Nathaniel A. Warne. 2018. "Should We All Be Scientists? Re-Thinking Laboratory Research as a Calling." *Science and Engineering Ethics* 24, no. 4: 1161–79.

Bird, Stephanie J. 2001. "Mentors, Advisors and Supervisors: Their Role in Teaching Responsible Research Conduct." *Science and Engineering Ethics* 7, no. 4: 455–68.

Caruana, Louis. 2006. *Science and Virtue: An Essay on the Impact of the Scientific Mentality on Moral Character*. Hampshire, UK: Ashgate.

CGS (Council of Graduate Schools). 2008. *Best Practices in Graduate Education for the Responsible Conduct of Research*. Washington, DC: Council of Graduate Schools.Chen, Jiin-Yu. 2015. "Virtue and the Scientist: Using Virtue Ethics to Examine Science's Ethical and Moral Challenges." *Science and Engineering Ethics* 21, no. 1: 75–94.

Chen, Jiin-Yu. 2016. "Research as Profession and Practice: Frameworks for Guiding the Responsible Conduct of Research." *Accountability in Research* 23, no. 6: 351–73.

Deming, Nicole, Kelly Fryer-Edwards, Denise Dudzinski, Helene Starks, Julie Culver, Elizabeth Hopley, Lynne Robins, and Wylie Burke. 2007. "Incorporating Principles and Practical Wisdom in Research Ethics Education: A Preliminary Study." *Academic Medicine* 82, no. 1: 18–23.

DuBois, James M., Debie A. Schilling, Elizabeth Heitman, Nicholas H. Steneck, and Alexander A. Kon. 2010. "Instruction in the Responsible Conduct of Research: An Inventory of Programs and Materials Within CTSAs." *Clinical and Translational Science* 3, no. 3: 109–11.

Georgia Tech. n.d. "Responsible Conduct of Research." http://rcr.gatech.edu. Accessed December 21, 2020.

Heitman, Elizabeth, and Ruth Ellen Bulger. 2005. "Assessing the Educational Literature in the Responsible Conduct of Research for Core Content." *Accountability in Research* 12, no. 3: 207–24.Kalichman, Michael. 2007. "Responding to Challenges in Educating for the Responsible Conduct of Research." *Academic Medicine* 82, no. 9: 870–75.

Kalichman, Michael. 2014. "Rescuing Responsible Conduct of Research (RCR) Education." *Accountability in Research* 21, no. 1: 68–83.

Kalichman, Michael, and Dena Plemmons. 2018. "Intervention to Promote Responsible Conduct of Research Mentoring." *Science and Engineering Ethics* 24: 699–725.

Kligyte, Vykinta, Richard T. Marcy, Ethan P. Waples, Sydney T. Sevier, Elaine S. Godfrey, Michael D. Mumford, and Dean F. Hougen. 2008. "Application of a Sensemaking Approach to Ethics Training in the Physical Sciences and Engineering." *Science and Engineering Ethics* 14, no. 2: 251–78.

Lang, James M. 2013. *Cheating Lessons: Learning from Academic Dishonesty*. Cambridge, MA: Harvard University Press.

Macfarlane, Bruce. 2009. *Researching with Integrity: The Ethics of Academic Enquiry*. New York: Routledge.

MacIntyre, Alasdair. 2007. *After Virtue: A Study in Moral Theory*. 3rd ed. Notre Dame, IN: University of Notre Dame Press.Mastroianni, Anna C., and Jeffrey P. Kahn. 1998. "The Importance of Expanding Current Training in the Responsible Conduct of Research." *Academic Medicine* 73, no. 12: 1249–54.

Mastroianni, Anna C., and Jeffrey P. Kahn. 1999. "Encouraging Accountability in Research: A Pilot Assessment of Training Efforts." *Accountability in Research* 7, no. 1: 85–100.

Michigan State University. n.d. "Research Integrity." https://grad.msu.edu/research-integrity. Accessed December 21, 2020.

Mumford, Michael D., Shane Connelly, Ryan P. Brown, Stephen T. Murphy, Jason H. Hill, Alison L. Antes, Ethan P. Waples, and Lynn D. Devenport. 2008. "A Sensemaking Approach to Ethics Training for Scientists: Preliminary Evidence of Training Effectiveness." *Ethics & Behavior* 18, no. 4: 315–39.

Nakamura, Jeanne, and Michael Condren. 2018. "A Systems Perspective on the Role Mentors Play in the Cultivation of Virtue." *Journal of Moral Education* 47, no. 3: 316–32.

NASEM (National Academy of Sciences, Engineering, and Medicine). 2017. *Fostering Integrity in Research*. Washington, DC: National Academies Press.

NIH (National Institutes of Health). 2009. "Update on the Requirement for Instruction in the Responsible Conduct of Research." https://grants.nih.gov/grants/guide/notice-files/not-od-10-019.html

NIH and ADAMHA (National Institutes of Health and Alcohol, Drug Abuse, and Mental Health Administration). 1989. "Requirement for Programs on the Responsible Conduct of Research in National Research Service Award Institutional Training Programs." In *National Institutes of Health Guide for Grants and Contracts*. Bethesda, MD: National Institutes of Health.

Nature Careers. 2016. "Scientific Virtue." *Nature* 532, no. 7595: 139.

NSF (National Science Foundation). 2009. "Responsible Conduct of Research." *U. S. Federal Register* 74, no. 160: 42126-42128. https://www.govinfo.gov/content/pkg/FR-2009-08-20/html/E9-19930.htm. Accessed May 21, 2021.

OU (University of Oklahoma) Graduate College. n.d. "Professional Ethics Training." http://www.ou.edu/gradcollege/funding/graduate-assistantship/professional-ethics-training. Accessed November 15, 2018.

Pellegrino, Edmund D. 1992. "Character and the Ethical Conduct of Research." *Accountability in Research* 2, no. 1: 1–11.

Pennock, Robert T. 2018. "Beyond Research Ethics: How Scientific Virtue Theory Reframes and Extends Responsible Conduct of Research." In *Cultivating Moral Character and Virtue in Professional Practice*, edited by D. Carr. London: Routledge. 166–177.

Pennock, Robert T., and Michael O'Rourke. 2017. "Developing a Scientific Virtue-Based Approach to Science Ethics Training." *Science and Engineering Ethics* 21, no. 1: 243–62.

Phillips, Trisha, Franchesca Nestor, Gillian Beach, and Elizabeth Heitman. 2018. "America Competes at 5 Years: An Analysis of Research-Intensive Universities' RCR Training Plans." *Science and Engineering Ethics* 24, no. 1: 227–49.

Plemmons, Dena K., Suzanne A. Brody, and Michael W. Kalichman. 2006. "Student Perceptions of the Effectiveness of Education in the Responsible Conduct of Research." *Science and Engineering Ethics* 12, no. 3: 571–82.

Plemmons, Dena K., and Michael Kalichman. 2018. "Mentoring for Responsible Research: The Creation of a Curriculum for Faculty to Teach RCR in the Research Environment." *Science and Engineering Ethics* 24, no. 1: 207–26.

Resnik, David B. 2012. "Ethical Virtues in Scientific Research." *Accountability in Research* 19, no. 6: 329–43.

Resnik, David B., and Gregg E. Dinse. 2012. "Do U.S. Research Institutions Meet or Exceed Federal Mandates for Instruction in Responsible Conduct of Research? A National Survey." *Academic Medicine* 89, no. 9: 1237–42.

Schmaling, Karen B., and Arthur W. Blume. 2009. "Ethics Instruction Increases Graduate Students' Responsible Conduct of Research Knowledge but Not Moral Reasoning." *Accountability in Research* 16, no. 5: 268–83.

Schmidt, Karen L., Laurel Yasko, Michael Green, Jane Alexander, and Christopher Ryan. 2014. "Evolution of an Innovative Approach to the Delivery of In-Person Training in the Responsible Conduct of Research." *Clinical and Translational Science* 7, no. 6: 512–15.

Steneck, Nicholas H., and Ruth Ellen Bulger. 2007. "The History, Purpose, and Future of Instruction in the Responsible Conduct of Research." *Academic Medicine* 82, no. 9: 829–34.

Titus, Sandra L., and Janice M. Ballou. 2014. "Ensuring PhD Development of Responsible Conduct of Research Behaviors: Who's Responsible?" *Science and Engineering Ethics* 20, no. 1: 221–35.

USDA (United States Department of Agriculture). 2018. "Responsible and Ethical Conduct of Research." https://nifa.usda.gov/responsible-and-ethical-conduct-research.

13

Virtue Ethics and the Social Responsibilities of Researchers*

Mark Bourgeois

Introduction

Traditional modes of ethics training for scientists and engineers have tended to stress the basic ethical obligations of conducting research with honesty and integrity. Thus, this is typically known as research integrity training, or training in the responsible conduct of research (RCR) (Forge 2008; Macrina 2005). But the products of science and engineering research impact society in increasingly profound ways. It is therefore apparent that major ethical issues closely connected to science and technology research are not captured, or even approached, by such training.

Likewise, in the self-concept of science itself, there is a dual picture. In one, perhaps dominant, tradition the scientist only seeks the truth about the world, leaving what to do with or about those truths—even when they may constitute powerful new technologies—entirely to others (Pielke 2007). In the other, often observable, tradition scientists and engineers are deeply invested in the social and ethical impacts of their work, and engage in efforts to steer them in productive directions. The efforts of the Manhattan Project scientists to see nuclear power used responsibly is a prime example, but it is hardly an isolated case (Jungk 1958).

As the powers conferred on us by science and technology research grow, and their impacts on society become increasingly conspicuous, it is past time that the latter picture is emphasized, and that ethics training for scientists and engineers evolve to embrace the social responsibilities related to research. Yet due to their amorphousness and variability, this is a far more challenging area for ethics pedagogy than is simple research integrity. How can such responsibilities even be understood such that training can be provided? Following on our experiences developing and deploying an advanced social-responsibilities-of-research program for doctoral science, technology, engineering, and mathematics (STEM) students at the University of Notre Dame, with support from the National Science Foundation, this chapter will argue that the virtue ethics

Mark Bourgeois, *Virtue Ethics and the Social Responsibilities of Researchers* In: *Science, Technology, and Virtues.*
Edited by: Emanuele Ratti and Thomas A. Stapleford, Oxford University Press. © Oxford University Press 2021.
DOI: 10.1093/oso/9780190081713.003.0014

tradition provides essential conceptual resources both to frame and then to provide training in the social responsibilities of research. Following this argument, the chapter will provide an overview of how these insights were actually applied in our program. It will conclude by describing some of the outcomes of that program.

Social Responsibilities in Research

Rapid progress in scientific and technological research holds profound implications for our pursuit of, not to mention our very conception of, the "good life," whether as individuals or as a society. These implications may be either highly positive or strongly negative. Yet most often they present a complex mix that we constantly struggle to balance. As this research continues to present us with increasingly powerful capabilities to manipulate the world and ourselves, the challenge to responsibly wield these newfound powers will only mount.

But these breakthroughs do not fall from the sky. Nor do they roll off an automated assembly line. Each one is the deliberate, hard-won work of dedicated groups of human beings. Scientists and engineers bring us these innovations. We must therefore ask after their role in the social and ethical challenges that these then generate. What responsibilities do innovators have for the effects of their innovations?

There are a number of reasons to pose this question. The first begins with the innovators themselves. It should be acknowledged that by and large the scientists and engineers who labor to produce these breakthroughs do so with the genuine aspiration that they contribute to solving problems and making our lives better in some way. Indeed, perhaps only this could account for the endless hours of toil and iteration typical of technical research at the highest levels. Thus, asking after the responsibilities of the researchers for the effects of their research is not an attempt to preemptively place blame or to foist new, expansive responsibilities onto already conscientious agents. Rather, it is to ask how we, as educators, might better equip and support researchers to fulfill their already high-minded intentions.

But this is not the only, or perhaps even best, reason to pose this question. Another appears when the perspective is shifted from the researchers to the rest of society. Those who produce an innovation should, prima facie, be in the best position to understand it. This implies that the first, best chance to shape the impact of an innovation lies with those who produced it in the first place. At the very least, it implies that these innovators need to be an integral part of any conversation regarding how to manage those impacts.

For these reasons and more (some of which will become apparent shortly), researchers in science and engineering theoretically have vital ethical responsibilities regarding the impacts of their research products on society. These responsibilities clearly extend beyond the laboratory door; that is to say, they extend beyond the proper conduct of the research itself, and into the realm in which that research finds application. And yet for all that, research ethics training as it is currently provided very rarely mirrors this scope. Instead, for a variety of reasons, traditional RCR training—the only kind of ethics education typically mandated for graduate students in the sciences—focuses almost entirely on compliance with regulations and basic ethics within the performance of the research itself. RCR thus concerns issues such as research misconduct, data management, peer review, and plagiarism (Shamoo and Resnik 2009). As important—indeed, essential—as such issues are to the success of the research enterprise, they do nothing to address what happens once a research program does in fact succeed and enters society. Conventional ethics training for young scientists today is thus arguably incomplete—perhaps profoundly so.

A broader model for ethics education for scientists and engineers would also incorporate a recognition of the potential social responsibilities of research (Bird 2014; Bourgeois 2017). Such training would empower students to take a more holistic view of their ethical role as researchers, and also provide them with the skills to meaningfully engage with the impacts and context of their research. What exactly this engagement might look like will vary a great deal depending on the nature of the research and the state of its development. But it could include things like informing the public or relevant stakeholders on imminent novel technological capabilities; ensuring that the results of policy-relevant research are clearly communicated to appropriate policymakers; engaging with researchers in other fields to ensure that synergistic contributions to social problems are made; learning more about the social context of an issue to better understand what kinds of research products would make a real difference; or simply helping the general public understand the results of research with strong societal implications (e.g., on public health or the environment). Of course, this long list merely gestures at the various sorts of social roles a researcher may decide they must play in order to maximize the positive impact of their research.

Ideally, such social and ethical considerations would be a routine, embedded part of being a researcher—not an afterthought or a diversion. Yet despite their being essential to the ultimate good of the research, virtually none of these roles are countenanced in conventional ethics training for scientists and engineers (National Academies 2009). This is in large part because they represent a distinct break with a traditional, more confined vision of the scope of ethical responsibility in STEM. Hence, there is a substantial and urgent need to develop new paradigms, programs, and pedagogies to expand ethics training for scientists and engineers so that they embrace a broader scope of social and ethical responsibilities. In this, virtue ethics will be seen to play an essential role.

Virtue Ethics in Social Responsibilities

This broader vision for ethics education faces a number of substantial obstacles, both conceptual and practical, to being realized. While the practical barriers (including things such as institutional inertia and a regulatory-compliance orientation) are largely beyond the scope of the present chapter, the conceptual ones are central because they highlight the unique value of taking a virtue ethics approach in this more expansive vision of ethics education. The three primary conceptual challenges concern the determination, embedding, and normalizing of social responsibilities:

1. *Determination.* How should these broad, seemingly nebulous social responsibilities be understood or defined?
2. *Embedding.* How can such considerations become fully integrated into the routine conduct of researchers?
3. *Normalizing.* How can these modes be established as a "new normal" so that they take effect at scale, rather than being idiosyncratic?

Let us consider each of these challenges in turn.

Determination

To begin with, it is clear that there can be no recipe, procedure, algorithm, or rule that can precisely determine what social responsibility looks like for a particular researcher or research area. One reason for this is that each research project is unique in terms of its potential social implications. The relevant ethical concerns for a new medical therapy are apt to be quite different than those for a new industrial process, and different again from more basic research with less foreseeable applications. These variations are far too numerous for any conceivable set of rules. Even seemingly similar projects within the same field may have quite different social implications depending on local context—for example, a therapy for a disease condition predominantly found in the developed world (such as diabetes) will involve very different social considerations from one predominantly found in the developing world (such as malaria). And this is but one dimension of difference; another may lie in the techniques the research involves. For example, due to its highly general scope of application, research on a therapy that utilizes gene-editing technology presents quite a different ethical complexion than one that uses conventional pharmaceutical approaches (and this remains so even if they are targeting the very same condition). In short, the variations in social context, technological implications, local research context, and more present too many dimensions of variability for any conceivable set of rules or guidelines to fruitfully address.

Moreover, there is very likely to be such a multiplicity of social considerations even within a single research project that not all of them can or should be pursued. Not only will time and resource constraints demand this, but the ability to address different considerations may create tensions among those considerations, even to the degree of being mutually exclusive. In such cases, even should a set of rules be able to pinpoint potential responsibilities, it would then need to also specify how to choose between them.

Finally, "social responsibilities" are not confined to the societal impacts of research, or even the interaction of society and research. The research or disciplinary community itself constitutes a social setting, and there may be issues to address here as well. For example, demographic underrepresentation of both women and racial minorities remains a stubborn problem in the sciences. Disciplinary or departmental culture may also present opportunities for improvement or for service. As they pertain to the context in which the research is actually performed, and thus constrain how it is conducted and by whom, these "in-house" issues may be at least as important to the ultimate social impact of science as the issues more immediately related to that impact.

In sum, as critical as the social responsibilities of researchers are, their complexity, multiplicity, and variety mean that they simply cannot be advanced through establishing a set of processes or enacting a set of rules.

This principle has profound implications for ethics pedagogy and the use of ethical theory in social responsibilities training programs. Traditional approaches to ethics education, in research ethics and in applied ethics more generally, tend to highlight deontology and consequentialism as the primary moral theories available to enable the student to engage in "ethical decision-making" (e.g., National Academies 2009). Of course, this framing is itself problematic. Ethical challenges rarely present themselves as prepackaged dilemmas, ready to be submitted to analysis. Indeed, one often needs experience and long reflection to develop the keen moral sensitivity that allows ethical tensions or inequalities to be recognized—perhaps even if these are profound or in plain view. Moreover, such a framing presumes that ethical problems are exclusively about preventing moral harms. As the earlier treatment of social responsibilities already indicates, at least as important is maximizing moral benefits. This is not a mere semantic difference. Particularly in scientific and technological contexts, where impacts can be far-reaching and difficult to predict, an ethic of moral aspiration, as opposed to one of simple noncommission, is crucial.

But the framing of ethics as decision-making is only half the difficulty here. Equally problematic is the suggestion of process-based ethical theories as the sole tools with which to approach ethical matters. Deontology and consequentialism are both salient and remarkably useful perspectives. Each captures something essential about the nature of morality. Surely making everyone better off is a primary goal of ethics, and just as surely people should be treated as ends in themselves. Famously, of course, these seemingly noncontroversial and self-evident principles can and do come into dramatic conflict with each other, such that only one or the other may be observed. But it is not that exclusivity that presents the most serious challenge. Rather, it is the fact that both moral theories are enacted as analytic processes—essentially, as algorithms. Consequentialism would have us first decide on a universal metric of utility, whether dollars or life-years, and then rank-order alternative courses of action based on their expected outcomes in those terms, taking the action that is expected to produce the highest net payoff. Deontology, according to the first formulation of Kant's categorical imperative, would have us first formulate the maxim under which a proposed action falls, and then determine whether or not that maxim is logically consistent when universalized. We are, of course, to act only according to those maxims that are. These consistent maxims, such as "do not lie" and "do not kill," thus become universally binding exceptionless rules that we must follow. For all their genuine differences, then, and for all their genuine insight, both consequentialism and deontology require us to first distinctly frame the moral problem and then to formulate alternative courses of available action, which are then analyzed according to a kind of formula.

The problem with this process when applied to social responsibilities in research should by now be apparent. This is a moral space, not unlike much of real life, which is resistant to cut-and-dried formulation and analysis. Should a given researcher invest their time engaging policymakers, or the general public? Should they tackle possible patterns of sexual harassment in their own department, or engage with researchers in a different discipline working on other aspects of a common problem? Or should they spend time getting to know more about who and what stands to be impacted by their research? Each alternative presents a complex set of trade-offs. Each represents noncomparable kinds of goods. And each has a deeply uncertain payoff. Here, the relatively formal ethical analysis of the traditional duo of consequentialism and deontology struggles to gain a foothold.

And just here is where the strengths of a virtue-ethics-based approach, drawing on the Aristotelian tradition, become apparent. For in virtue ethics, the cultivation of moral judgment and practical wisdom replaces rules and rubrics (Aristotle 2014; Schwartz and Sharpe 2010). A very brief review will serve to make this synergy clearer. At its root, virtue ethics, as first articulated in the *Nicomachean Ethics*, is about a vision of the good life for human beings—of what constitutes flourishing (or *eudaimonia*) for us as individuals, as communities, and even as a species. For us as individuals, what that flourishing looks like in its specifics will depend on our particular context—on the society we find ourselves in and on what role we aspire to play within it (MacIntyre 2007). But while the specifics vary, the pattern remains consistent. Flourishing requires the attainment of various kinds of goods, and these in turn require the cultivation of certain moral habits: the virtues that give the approach its name. Virtues are dispositions, skills, or habits that tend to enable success in the attainment of some unique good, one that is characteristic of a particular type of activity. Virtues for Aristotle are, famously, means between vices—one of excess and one of deficiency (Aristotle 2014, book 2). For example, in war courage is a virtue, and courage is found partway between the vices of cowardice (an excess of fear) and rashness (a deficiency of fear).

As will become important in the next section, these virtues are indeed tendencies or habits. Insofar as this is the case, they are established dispositions in the individual that do not require fresh analysis and deliberation each time that they are called for. The virtuous soldier need not deliberate or remind themselves to have courage each time it is required. They simply are courageous by default. And yet this is not a thoughtless process, either. It is not as if the soldier attains the relevant virtues once and need never think about what they do again—indeed, it is quite the opposite. For these virtues are never self-explanatory. Neither are they singular. Rather, while individually and in simple cases they function nearly automatically once acquired, the world is rarely so simple. As often as not, virtues require thoughtful application. This is why Aristotle not only identified a variety of virtues appropriate to various settings but also identified a "chief" virtue, phronesis.

Phronesis (the Greek word is sometimes translated as "prudence") is, essentially, moral judgment or "practical wisdom." It is knowing which virtue to deploy at what time, and to what degree. It is also knowing how to balance multiple virtues when not all can be exercised at once or when they may even be in conflict with one another (Aristotle 2014, book 6; Schwartz and Sharpe 2010). The courage to simply face battle is one thing. Whether to risk the mission to rescue an injured comrade is another. Whether to fire on an unidentified party that may be enemy or may be civilian is yet another. All of these require judgment based on experience and a sense of priorities. Each is a deeply moral decision and yet at the same time also a practical decision about what will best achieve the goals that have been set. Each requires skillful deliberation, but also quick and decisive action. Each such choice is unique, conditioned by its own distinct set of circumstances and uncertainties. Therefore, the decision cannot be made automatically, solely from habit. But without having first cultivated an ensemble of the relevant virtuous dispositions (such as courage, compassion, and dedication)—and without having a settled vision of the goal or good of the activity—consistently making such fraught decisions successfully would be nearly impossible.

Fortunately, the pressures of being a socially responsible researcher are far less extreme. But the same principles apply. Where moral rubrics or preset lists of defined responsibilities fail, the cultivation of virtues, attention to goals and purposes, and the development of moral judgment are called for. Only these will equip emerging researchers to decide for themselves—not all at once, but over the course of their development and maturation as researchers—what social responsibilities they are called to engage with, and how. This will inevitably depend on the nature of their discipline, their available resources, the stakes of their particular research project, and their own skills, interests, and priorities. But the virtue ethics framework provides the flexibility to take all this into account. In this respect, a virtue ethics approach to training for social responsibilities is a natural, perhaps essential, fit. But this is so for other reasons as well.

Embedding

The second major conceptual challenge in social-responsibilities-for-researchers training is how to frame and promote these broader considerations as routine parts of engaging in research, rather than as optional, supererogatory activities that can just as easily be neglected. Here, too, a virtue ethics approach provides essential resources.

The first hurdle to embedding is the perception that social concerns and activities are not themselves properly a part of research (or science) but instead extracurricular activities that researchers may choose to engage in, but at the expense of doing the "real" work of research. On this traditional view, the researcher is responsible for the ethical and accurate conduct of research in their own lab, and perhaps partly, in their broader role as faculty member, for the culture of their local department. But concern for public impact, policymaking, or engagement with stakeholders constitutes a distraction from their true remit as researchers. This view presumes that the role of the researcher is simply to generate new knowledge and that it is up to other members of society to decide what, if anything, to do with it.

There are, of course, a number of ways in which this view is already tenuous in principle. For example, it would seem implausible that an environmental researcher who uncovers a serious risk would be indifferent as to how and whether that risk is acted on. It seems more likely that someone who has dedicated their professional life to researching environmental change does so with the conviction that their work should positively contribute to the preservation of the environment. In other cases, the distance between discovery and application may be so short as to be imperceptible. Researchers who, for example, develop a novel energy technology can hardly claim to have no responsibility for the impacts when that technology is actually implemented. Nonetheless, we might imagine some researchers to claim a distinction between their duty and role as a researcher per se and their interests and role as a member of society or as a citizen.

So, what is the proper role of the researcher, and will engaging in social responsibility activities help them to fulfill it, or distract them from it? This is to ask whether doing so will increase the success of their research. And this, of course, depends on how that success is defined. It is unlikely to expedite the research process, or to increase its scientific accuracy or reproducibility. It is equally unlikely to increase the number of scientific publications generated. If these are defined as the true goals of research, then social responsibilities would indeed seem to be a distraction.

Engaging in social responsibility may, however, increase the value of the research that is produced—both because research that would have been done anyway will be more likely to reach those who can use it in a usable form, and also because it may provide feedback that redirects the research program to still more valuable products. If the goal of the research is connected to the value of its product for society rather than to more traditional metrics of productivity, then engagement with social responsibilities may in fact substantially advance the success of research.

This, however, leads to a second iteration of the embedding problem: the incentive structure of most research positions. This second manifestation is both a cause and an effect of the first. Generically speaking, incentive structures are erected to encourage the attainment of some good. Thus, salespersons are paid on commission for the amount of product they sell. But the incentives themselves are merely proxies. The commission is not the sale, much less the revenue for the company, the jobs for the manufacturing employees, or the satisfaction of the customer need. These are the goods that the commission is truly meant to bring about. But of course from the salesperson's perspective, forced to make a living from them, the commission may very well become the good—the only goal. Researchers are not immune to this effect.

Whether academic, industrial, or governmental, researchers are generally rewarded based on proxies for the good genuinely sought. These include rewards (such as hiring, tenure, promotion, or raises) for things such as publications or patents, based on the premise that these products reflect new and presumably valuable research findings. There are rarely rewards—at least not directly—for ensuring that those findings are communicated to policymakers, or that they address a need that stakeholders in the field have articulated, or that the public understands and welcomes them. These crucial qualitative aspects more closely represent the true goods sought, but they elude the established incentive structures.

Reform of those incentive structures is in most cases unlikely. Yet even if they were more susceptible to reform, prescriptive processes and rules would not suffice to promote genuine social responsibilities. As has already been established, there are simply too many considerations and permutations. But beyond this, new and well-intentioned incentives for social responsibilities, such as stakeholder engagement, could even conceivably serve to stifle genuine concern for social responsibilities, or to channel them in unproductive directions. Incentives require measurement, and once measured, those metrics become new proxies. As Goodhart's law states, "When a measure becomes a target, it ceases to be a good measure" (Hoskins 1996; Muller 2018). In other words, it is unlikely that any reward structure for social responsibility would actually succeed in fostering a genuine attitude of responsibility, simply because it is a reward structure.

The standpoint of social responsibilities asks researchers to reconsider, or at least recontextualize, the minimal-responsibility view of science. Ultimately, the question is: is the good of the research in simply arriving at an answer, or is it in the good that can be done with that answer? So long as the two are thought to be inextricable, the answer matters little. The latter will inevitably follow from the former. And if these are handled by separate parties in society, that, too, would ultimately matter little in principle. But this is arguably not the case. Research that fails to understand the social context that requires it will fail to make a difference there. Policy-relevant research that is not communicated in usable form will fall through the cracks. And innovations that are feared or not well understood by the public will be rejected, no matter what benefits they may offer (van den Hoven 2013).

This leaves the researcher with two choices: either confine oneself to the narrow vision of the researcher as knowledge producer, or begin to engage with the potential social and ethical ramifications of the research. Choosing the latter is something that can be persuasively argued for, but it cannot be compelled. Yet without training in social responsibilities, the choice itself will likely remain invisible to the researcher. Importantly, this training must include more than an articulation of the choice; it must also include an introduction to at least some of the skills that would enable such responsibilities to actually be undertaken— otherwise the choice remains merely theoretical.

In the end, for broader social and ethical considerations to become routine, especially in the absence of (if not the opposition of) incentive structures, there seems only one viable solution: if they are to become a part of the character of research, they must become a part of the character of the researcher. That is, concern for these aspects must become embedded as a moral habit—as a virtue—in the individual researcher. Only then will their consideration become embedded as part of the research process.

Normalizing

The final conceptual challenge to social responsibility framing and pedagogy is how to establish this approach to research as the default orientation—not merely for an individual researcher but for an entire discipline. The answer arrived at earlier has one obvious problem: if embedding attitudes of social responsibility occurs on the level of an individual's moral development, then the growth in the prevalence of this attitude will happen only one researcher at a time. This might, in our rapidly changing world, seem to be a hopelessly slow progression. But it need not be. After all, online memes and communicable diseases spread only one person at a time as well—and these can reach epidemic proportions very quickly. The key is transmissibility. This, too, is an idea built into the virtue ethics approach, in the form of the moral exemplar.

The approach of virtue ethics asks much from its adherents. Without rules or rubrics, one is expected to exercise sound judgment, cultivate productive virtues, and develop refined practical wisdom. How, it is fair to ask, is one expected to do all this? The short answer is that one does not do it alone (Aristotle 2014, books 2, 8, and 9; MacIntyre 2007). Again unlike deontology or consequentialism, which can be adopted and deployed in private by an individual, virtue ethics is very much a team sport. It is not too hard to see why this should be. After all, only a community could decide the goals of a communal activity, or determine which behavior traits (i.e., virtues) have tended to lead to it (MacIntyre 2007).

Because virtue ethics is communal, methods for inculcating others into the community are built into the approach. These methods escalate as the experience, skill, and sophistication of the learner build. To begin with, didactic instruction will provide the initial foundation, as it does with children being raised. Soon they will move from receiving instruction to receiving feedback as they attempt to put their lessons into practice. But as the learner grows in skill and perception, they will soon reach a point where they have absorbed most of what can be framed as simple instructions or direct feedback on simple tasks. This is where the real moral development begins. From this point, rather than being told what to do, they must learn for themselves via long experience and practice in the community, as they would with any skill (Russell 2015). But in support of that lifelong task, they should also closely observe morally mature practitioners so that they can be modeled. These are the moral exemplars.

Of course, students may not have the opportunity to observe the most notable exemplars in person. But not all exemplars need to be seen in person. Indeed, often the most compelling examples come from history. The story of the Scientists' Movement, for instance, is a powerful example of researchers squarely facing the momentous social impact of their own nuclear research and determinedly steering it to a more socially productive end (in this case, by establishing civilian, rather than exclusively military, control over nuclear technology) (Jungk 1958). Jonas Salk's refusal to patent the polio vaccine, the story of courageous whistleblowers in the tobacco industry, and the biotechnology research community's decision to pause early recombinant genetic research in order to address ethical concerns all serve as other instructive examples of researchers taking social responsibility. Nonetheless, local examples, while they may be at a smaller scale, are equally important, for they convey that such responsibilities are not reserved for those whose names or stories were considered exceptional enough to be recorded by history.

Students will almost certainly be unable to observe exemplars engaging the precise dilemmas and situations that the learners themselves face, or will face. But the lessons of these examples lie not in the details of the scenarios but in the virtues of the practitioners who successfully navigated them. Study of the characters of those examples that they can observe will allow students to begin modeling those traits in themselves—whether that be the courage of speaking truth to power or the dogged determination not to allow one's work to be misused.

Training in social responsibilities for young scientists and engineers must thus do two things. First, it must offer them real examples of researchers engaging in social responsibility, in ways large and small, current and historical, local and global. Second, it must clearly communicate to them that if they choose to engage with these responsibilities, in doing so they will be setting themselves up as moral exemplars for others.

If more students are able to convincingly demonstrate the virtues associated with the social responsibilities of research, more of their peers, their future colleagues, and eventually their own students will come to recognize these as essential to the life and work of research. And as this happens, gradually a new normal may be established. On the other hand, perhaps fully embracing these values will remain the province of only those sufficiently motivated by the higher aims of research. Even in this latter case, however, the need for more such exemplars is vital. In any case, the moral modeling at the core of virtue ethics constitutes one of the few plausible tools to raise the moral consciousness of an entire community.

The SRR Program at the University of Notre Dame

Motivated by the concerns discussed here, in 2013 the Reilly Center for Science, Technology and Values at the University of Notre Dame proposed a virtue-ethics-inspired training program on social responsibilities to the National Science Foundation (NSF) under its Ethics Education for Scientists and Engineers (EESE) program. This proposal, authored by professor of philosophy and then Reilly Center director Don Howard as well as then Reilly Center assistant director for research Melinda Gormley, was awarded. The author of the present chapter was subsequently brought on to design and execute the resulting program.

The proposal requested support for three iterations of a year-long training program in social responsibilities. The training itself was to consist of a week-long "boot camp" at the start of summer, regular meetings during the academic year, and, centrally, a social responsibility project of the student's own design undertaken with the support of a mentor (either Howard, Gormley, or the author). The cohorts were to be around fifteen students in size, consisting of second- or third-year PhD students in STEM disciplines—broadly interpreted to include natural science, social science, and engineering. The program years were chosen to roughly coincide with the transition from coursework to research work. In order to incentivize participation, students were to receive a stipend of $1,400 as well as a small ($200) budget for their projects. Beyond incentivizing, the stipend was also intended to communicate to the student's advisor and lab that we valued the student's time and effort, and preempt the perception that we were drawing on these at no cost. The actual training program came to be known as SRR, for Social Responsibilities of Researchers, and stayed true to the proposed pattern. Further details of the program's goals, pedagogy, and logistics, as well as a number of the student projects that arose from it, are described in what follows.

Goals and Framework

One of the earliest challenges in the work of developing the programming (and its assessment) was creating a framework to give structure to the rough concept of social responsibilities. While, as discussed at length earlier, it is impossible to provide a single rubric or a defined set of rules to follow to be "socially responsible," some articulation of the fundamental components was necessary to create the program. One way of arriving at this is to step through a general case of a researcher who desires to exercise social responsibility for research impacts.

For example, imagine a biologist researching antibiotic resistance who wants to know how to steer that research for maximal social benefit (the actual case for a student in the first cohort of SRR). To begin with, that researcher needs an understanding of the issues and contexts that their work may enter into should it succeed and find application outside of the lab. One obvious area to look at is medical care, since that is where antibiotic resistance actually poses its risks. But in this case, that turned out not to be the most salient area, for the bulk of antibiotics in this country are actually used in livestock, and it is in livestock that bacterial resistance tends to develop. Only from there does it go on to present a threat to human health. This insight alone is crucial for understanding the paths available through which to make meaningful scientific contributions; yet the student's training in the biology of antibiotic resistance alone had not provided it.

With a basic understanding of the real-world context and how their own work might interact with it, the next step for the student might be to engage the people relevant to that context to learn more about it and to open the lines of communication regarding the research. In this particular scenario, that may mean speaking to policymakers tasked with setting rules for the provision of antibiotics to farm animals, and, eventually, perhaps even to the farmers themselves. This requires skill in communicating science to nonscientists and policymakers, both to explain the researcher's own work and to learn more about the practical constraints involved.

The purpose of this interaction is an ongoing incorporation of social insights, concerns, and impacts back into the research project. This could take many different forms, such as modifying the goal of the research to address a practical problem encountered in the field, or changing research methods to take account of new information or newly available data sources. Perhaps no changes will be necessary, but the engaged researcher will know this, too—and know when and if it changes.

Throughout this process the researcher must evaluate their own moral values and priorities and learn to bring them to bear in the work, for it is ultimately these that drive the entire activity. Likewise, the researcher must (like the rest of us) be able to critically monitor their own motives and assess their success from time to time. Finally, any researcher who values engaging in this fashion must be able to model these values and behaviors for others in their research community, including junior personnel and students. For the reasons explained earlier, this is how such values and skills will be spread.

Putting this sequence of activities together yields a set of six goals for social responsibility training to aim to cultivate:

1. *Understand context.* Researchers should become more sophisticated in their understanding of the broad social and political context their work products may enter into. This goal is first because it is the most fundamental— a prerequisite for all others. Social, financial, political, geographic,

ethical, religious, and professional contexts could all be materially relevant in shaping research impacts. We can't make students experts in all these areas, but they should recognize how some might condition their work.

2. *Perceive impacts.* Researchers should learn to think about the impacts their work could have in these contexts, whether in direct consequences or in opportunities to contribute to social needs. The point of recognizing and studying context is to imagine how their research might change it.

3. *Engage others.* Researchers should learn to communicate and interact effectively with the general public, with stakeholders in their research, with cross-disciplinary collaborators, or with policymakers. This engagement and communication should be bidirectional. Not only is explaining the science and its relevance important, but better interaction with stakeholders can lead to improved understandings of context for the researcher. Because impacts often cross disciplinary boundaries, interdisciplinary collaborations may also be critical, and engagement skills are equally important there.

4. *Adapt research.* Researchers should learn to consider how to adapt their research as appropriate to enhance its social relevance and impact, including modifying its direction or its conduct. Ultimately, engagement and perceiving impacts may only be valuable if one is prepared to modify research plans in response. This might range from tweaking the goals of a research project to undertaking an entirely new one.

5. *Self-reflect.* Researchers should become more cognizant of their own values, motivations, and ethical orientation, so that they can maintain focus and defend against complacency. Moral psychology research shows that we all need to learn to be critical of our motivations when necessary. Equally, articulating our core values allows us to defend our decisions and adhere to them more consistently.

6. *Model and mentor.* Researchers should become comfortable acting as a mentor and ethical model, helping to change the culture of their lab, field, or discipline by serving as a model of ethical engagement. Social responsibility is both individual and disciplinary. Modeling and mentoring are core skills to move from one to the other.

These six goals, collectively, have become the working definition of SRR training at Notre Dame, guiding not just SRR but two subsequent graduate training programs as well (Bourgeois 2017). While the SRR training program itself concluded with the third of three cohorts finishing in 2018, social responsibilities training is currently ongoing in an internally supported cohort-based fellowship program in the Graduate School that also incorporates ethical leadership development. That program is known as LASER, for Leadership Advancing Socially Engaged Research.[1]

SRR Program Logistics and Pedagogy

Because the SRR program was a cohort-based year-long program that required a significant ongoing commitment from each fellow, recruitment and selection were a crucial first step for the program each year. After widespread recruitment efforts, applications typically exceeded program spaces by about two to one, confirming that many students do see value in holistic ethics-oriented training programs such as this. Applications consisted of letters of intent, letters of recommendation or permission from their advisors (or graduate program directors for those not yet assigned an advisor), and a CV or resume. Applicants were requested to describe their intended social engagement project in their application letter, or at least to identify the area they were interested in pursuing for their project. Demographic and disciplinary diversity of each cohort overall were also factors in selection.

Following admittance to the program early in the spring semester, the cohort was provided with a selection of preparatory reading material to complete before the week-long boot camp took place at the very beginning of summer. The boot camp was our main opportunity to deliver formal classroom training; however, the sessions were made as interactive as possible. The boot camp was five consecutive full days, from 8 a.m. to 5 p.m., Monday through Friday. Aside from a field trip in the middle of that time, all sessions were held in a single classroom. This concentration in terms of both time and space, which gave rise to the nickname "boot camp," was a deliberate pedagogical choice as well as a logistical convenience. The pedagogical intent was to help forge the diverse interdisciplinary group into a cohesive community and to allow for extended, ongoing conversations across the whole week, without the usual interruptions and distractions. The logistical factor was that many students, especially in the social sciences, travel for the summer to do fieldwork or other kinds of research, while during the academic year it is impossible to obtain uninterrupted time of such duration. By choosing the first week of summer for the boot camp, we avoided academic year pressures while still enabling students to travel for the rest of the summer.

The content of the boot camp training itself largely fell into four major categories: ethics, exemplars, communication, and project development. The ethics content was devoted to covering the standard philosophical approaches to applied ethics, including virtue ethics. While the program as a whole is consciously founded on the virtue ethics approach, there is still much value in learning to think analytically and philosophically about ethics in general, particularly when students are coming from disciplines, such as science and engineering, that typically do not incorporate much humanities work in their curricula. Since virtue ethics is communicated implicitly throughout the program, the explicit content was limited to a conceptual and historical overview. For the frameworks of deontology and consequentialism, a case study in pharmaceutical clinical trials in the developing world was closely analyzed. In this case, the consequentialist good that this drug study would do in terms of saving lives had to be weighed against the undeniable fact that the subjects who enrolled in it would be used as a means to an end: that of testing the drug so that it could be approved for sale in the developed world (Bourgeois 2012). This case never failed to generate passionate debate and discussion, which often continued even after boot camp ended. (Indeed, one student project, described later, was at least partly inspired by this case study.) Finally, an exercise on formal codes of ethics asked students to discover and then to critique the code of ethics, if any, that governs their own discipline. The lacunae in these codes can be particularly instructive in the context of social responsibilities.

The content on exemplars, reflecting the analysis described earlier, came in two very different forms: historical case studies and in-person local exemplars. Case studies were selected to reflect each domain represented in SRR: natural science, social science, and engineering. The centerpiece of the case studies was an example from natural science, the Scientists' Movement of the 1940s and 1950s. Given the scale of the stakes and the famous figures involved, including Albert Einstein, this is perhaps the preeminent example of scientists both producing and taking serious moral responsibility for the social and ethical impacts of their research. Among the notable legacies of this movement is civilian, rather than exclusively military, control over nuclear technology in the United States. This unit was taught by Howard, the primary author and principal investigator of the proposal for SRR, who is also an expert in the history and philosophy of quantum mechanics as well as a former editor of the Einstein Papers Project. The social science case study also involved the American military, though in a more contemporary context. This case study, which takes the form of an interactive role-play class exercise, concerns the Human Terrain Systems project during the Iraq and Afghanistan conflicts.[2] The idea of this project was not simply to consult with social scientists—principally anthropologists—but to actually embed them in combat units in the field in order to improve the cultural literacy of the units and reduce tensions as well as casualties. Even before the first combat deaths of anthropologists in the field, the American Anthropological Association condemned this use of researchers as an "unacceptable application of anthropological expertise." The role-play involves balancing the aims of the military and the chance to reduce casualties with the proper role of anthropologists. Finally, for the engineering case study, students explored the geoengineering SPICE (Stratospheric Particle Injection for Climate Engineering) project in the United Kingdom.[3] In this project, scientists sought to do early field experiments of the atmospheric dispersion of particulate aerosols. In one of the main versions of proposed climate engineering, these aerosols would be used at scale to block a fraction of incoming solar energy, and so cool the planet. The SPICE project was to involve an extremely small-scale release of the particulates from a weather balloon, in order to study their dispersal patterns and use the results to refine computational models. However, when news of the experiment became public, there was a swift and severe backlash, and the experiment was cancelled. Notably, this objection was based not on the safety of the experiment itself but on the moral hazard that such work represents. The claim was that if such experiments proceed, the mere prospect of climate engineering would cause it to be portrayed as a viable alternative to reducing carbon emissions. In other words, experiments such as SPICE, however preliminary and necessary, would effectively take climate engineering from a fringe idea to a serious area of research, and perhaps provide yet another reason to delay the retirement of fossil fuels. In responding to the case, students must thus balance the good that such research aims at with the potential social consequences of its very conduct.

The local exemplars were likewise drawn from across the disciplines of natural science, social science, and engineering. These consisted of faculty members here at the University of Notre Dame whose work involves sustained outreach and engagement with stakeholders in their research. Each of these researchers gave a short talk describing their research, the way it integrally involves social engagement and a consideration of ethics, and how they came to do this work. But perhaps the greater value was in the informal, open-ended question-and-answer session that followed these talks, where students could begin to see why and how these researchers did what they did, and the nature of the personal commitment this involved. Researchers appearing at SRR boot camps across the three years of its operations included Professor Jen Tank of the Biology Department, who works with farmers across the Midwest to study crop runoff; Professor Marya Lieberman of the Chemistry Department, who developed paper test strips to detect counterfeit pharmaceuticals in the developing world; and Professor Darcia Narvaez of the Psychology Department, who publishes popular pieces on child-rearing, empathy, and moral development related to her research. Each of these researchers has also sent students from their own labs to participate in SRR. Between the historical case studies and the in-person contact with engaged researchers right here on campus, students are provided with a broad spectrum of moral exemplars and precedents for ethically and socially engaged research, taking the concept from aspirational theory to concrete reality.

Because communication skills are central to social engagement, these skills were a third major focus of the boot camp. Science communication skills were taught and practiced following the "message box" technique in Marcia Baron's book *Escape from the Ivory Tower* (Baron 2010). Personal presentation skills were practiced through *Shark Tank*–like "flash pitches" of the student's proposed project. And popular writing skills were taught and practiced with blog posts. Other components of the boot camp were dedicated to helping students develop and refine the plans for their own social engagement projects. This included a lecture and workshop on applying design thinking, as well as a survey of support resources and opportunities for engagement that the campus already offers. Notre Dame received a Carnegie Foundation classification for community engagement in 2011, and much of that work occurs through the Center for Social Concerns, a member of which presented at each boot camp.[4] Finally, in the middle of boot camp came a field trip to a local site where research and public impact meet. In the first two years, this took the form of a visit to the "smart sewers" firm emNet.[5] Drawing on biosensor engineering research conducted at Notre Dame, initially for military and homeland security applications, emNet was founded on the idea that this same sensor technology could be used to monitor sewer flows. The data that these sensors collect allows the sewers to be managed to prevent overflows during heavy rainfall, maximizing their existing capacity without costly investments to expand the system. In the final year of SRR, the field trip was instead to the office of the City of South Bend's chief innovation officer, Santiago Garces. This office, one of the first of its kind in the country, was created by Mayor Pete Buttigieg to leverage data and innovative technology for civic good. This meeting provided SRR students with a chance to see firsthand the myriad ways that research like theirs could contribute to addressing serious social problems, on issues ranging from infrastructure to policing.

At the conclusion of the boot camp each year, students in the SRR cohort have the rest of their summers to pursue their regular research while refining their project plans. When the academic year resumes, the cohort and the mentors meet as a group about every three weeks. These academic year sessions took place in the early evening and lasted for two hours. In the fall these sessions were primarily dedicated to providing additional skills and perspectives to follow on from the boot camp. These were informed by surveys at the end of boot camp asking what additional topics students would like to see. This resulted in sessions on the public perception of expertise and additional training on communication skills, including via social media. But the majority of the academic-year sessions were dedicated to student presentations on the progress of their engagement projects, modeled after a design review in industry, where cross-disciplinary teams provide feedback on evolving designs. This provided an opportunity for feedback from peers as well as mentors, and offered a chance for the students to show off what they had already accomplished. For while the training we provided was extensive and essential, much of the real growth took place via the students' own projects. Consequently, it is worth examining some of these.

Example Student Projects

While the SRR program, as described earlier, took a cohort-based approach and also involved as much in-person meeting as was feasible over the course of a year, the individual social engagement project that each student undertook on their own time still represented the largest investment of their time and effort. As it was the most self-directed aspect of the program, it is also where a substantial amount of growth and learning is likely to have occurred. Given the breadth of social responsibilities outlined previously, and given the personal and disciplinary diversity of the student fellows, these engagement projects not surprisingly ran the gamut in terms of form and content. While it is not feasible to catalog here each of the more than forty-five projects undertaken, it will be valuable to describe a few of the more exemplary ones to provide a sense of how profound their effect could be.

There is perhaps no better example of research with ethical impacts than the field of peace studies. Nonetheless, given the deeply interdisciplinary nature of this research, and the fact that by definition it is conducted in conflict zones, this type of research also poses acute ethical risks. Inspired by these issues, and prompted in part by the debate over the international clinical trial case study during boot camp, an anthropology and peace studies PhD student in the first cohort of SRR decided to author a journal article describing the unique ethical pressures of conflict research, highlighting the lack of existing resources with which to describe or address these, and proposing a simple conceptual framework as a possible starting point. The article, "Reflexivity, Responsibility and Reciprocity: Guiding Principles for Ethical Peace Research," was published in the *International Journal of Conflict Engagement and Resolution* in 2016 (Lederach 2016). As the abstract puts it, "The application of peace research to settings of violent conflict requires careful attention to the ethical dimensions of scholarship; yet, discussions about the ethics of peace research remain underdeveloped. This article addresses a critical gap in the literature, outlining a framework for ethical peace research broadly encompassed in three guiding principles: responsibility, reciprocity and reflexivity." Notably, these three principles were chosen in order to "offer a flexible framework" for research in a field where no single research method prevails, and where ethical codes from each individual discipline represented in the field (such as anthropology or sociology) would present a conflicted, piecemeal approach at best. Moreover, these principles are explicitly intended to capture not only the ethics of research practice but the social impacts and motivations of the research as well. As such, it is an ideal example of the embrace of social responsibility in research.

In the second cohort of SRR, a biology PhD student was studying mosquitoes and other insects as disease vectors, particularly of malaria. At this point in time, mid-2016, the Zika virus scare was in full swing, and confusion and scientific imprecision, if not misinformation, were rampant in the popular news media. This student felt that part of her social responsibilities as a researcher was to help improve the public's understanding of the science around insect-borne disease, especially as it related to highly publicized epidemics. Moreover, she wanted to find a way to share with a wider audience the fascination and thrill she felt as a researcher. Because Zika poses a particular threat to pregnant women, her first idea for a social engagement project was to partner with local women's healthcare providers to act as a resource for their patients. When this proved too narrow an audience and too logistically complex, the idea of a general podcast was hit upon. Thus, the podcast that came to be known as *Tiny Vampires* was born.[6] By its own description, this is "a show about insects that transmit disease and the scientists that are fighting them. Each episode is guided by questions sent in by listeners. The question is answered with background information and the story of how scientists were able to shine a light on that particular mystery." What is particularly unique about the approach of the show is that it first solicits questions from listeners (rather than telling them what it thinks they should want to know), and then attempts to answer those questions by examining, explaining, and translating specific scientific papers that address the question (rather than simply telling listeners what the researcher knows, or what "science" says). This enables the listener to understand not just the scientific facts but something of the real-world process of scientific inquiry, including its funding and publication. This has proven a successful formula. More than three years and thirty episodes later, it remains a thriving podcast with an array of distribution channels, partners, and sponsors. Moreover, this work directly led to this student's current job as a content developer for the Field Museum of Chicago's YouTube channel, The Brain Scoop.[7] The student described this as her "dream job," but it is one she would never have pursued if she had not had the chance to explore her social responsibilities as a researcher.

Other times, the SRR engagement project was so closely related to the student's research and uncovered such compelling resources or needs that it actually changed the course of their doctoral dissertation. Such was the case with a political science and peace studies PhD student in the second cohort. She was already doing research on women's rights in Africa, and her SRR social engagement project connected her with a local South Bend civic group called Malawi Matters.[8] This group, composed largely of Malawian immigrants, is focused on HIV and AIDS education in that country. This organization presented such a robust set of contacts and infrastructure that the student redirected the focus of her research to take advantage of them. In doing so, she also helped to build out the organization's programming for women and girls, in a course called "Equipping Women—Empowering Girls." For her dissertation research, she ultimately conducted a survey measuring the degree to which women across that country were aware of the rights they already enjoyed under their constitution— but which remained invisible if the women did not realize they had them. Given the low rates of literacy and the lack of telecommunication and transportation infrastructure, such a laborious paper-based survey and educational programs would not have been possible without the local resources and connections of the Malawi Matters organization. The results of this research are being used to not only combat HIV/AIDS and improve the plight of the women of Malawi but also to stem child marriage there.

Other engagement projects included the initiation of a Notre Dame chapter of the Science Policy Initiative, a group dedicated to connecting young researchers to policy issues and policymakers; a paper and presentation on the ethical and privacy issues of high-performance distributed computing research (Kremer-Herman 2017); the hosting of a science "boot camp" for middle-schoolers in Tanzania, and another in the United Arab Emirates; and a number of K-12 education interventions in the South Bend area.[9] In addition to projects connected to research impacts or education, some students chose to pursue projects related to their local department.

Conclusion

Being a researcher in any field of science, engineering, or social science involves certain baseline, non-negotiable ethical responsibilities. These include respecting the moral and legal rights of research subjects, conducting research with integrity, and honestly reporting findings. Much could be improved in the way that researchers are trained to undertake these responsibilities, but the training that most graduate students are required to have at least succeeds in articulating and emphasizing these. Yet immediately beyond these minimal obligations lies a wide vista of additional ethical considerations. And it is these considerations— more so even than the basic ones—that can affect how that research makes a difference in the real world. Yet not only do we neglect to train for these in any systematic fashion, but we often fail to even acknowledge them. At worst, we pretend that these responsibilities apply not to the researchers who produce the findings but only to other members of society.

None of this is to say that such responsibilities are not routinely acknowledged, embraced, and communicated by thoughtful PIs and researchers in every discipline and institution. But not every student will be fortunate enough to be trained by such a researcher, and many may complete their entire doctoral degree without such exposure. One barrier to more systematically approaching these responsibilities in the education of researchers has been the lack of a way to conceptually frame them, and thus to train for them. The virtue ethics tradition presents a rich set of resources with which to do just this, and the SRR program at the University of Notre Dame demonstrates one way in which this can be accomplished. Our experience with this program has convinced us that providing the opportunity to explore social responsibilities in a structured way should itself be part of the responsibilities of graduate educators.

Notes

* I would like to thank Don Howard: first, for his leadership in and authorship of the NSF EESE proposal for the SRR program; second, for his constant friendship and mentorship; and finally for setting an example of the truly socially engaged researcher. My thanks also to Melinda Gormley as co-author and co-PI of the original NSF EESE proposal. This material is based upon work supported by the National Science Foundation under award nos. 1338652 (SRR) and 1449469 (EL-STEM). Any opinions, findings, and conclusions or recommendations expressed in this material are those of the author(s) and do not necessarily reflect the views of the National Science Foundation.
1. University of Notre Dame Graduate School, "Leadership Advancing Socially Engaged. Research (LASER)," https://graduateschool.nd.edu/graduate-training/leadership/laser/.
2. "Human Terrain System," Wikipedia, https://en.wikipedia.org/wiki/Human_Terrain_System.
3. Stratospheric Particle Injection for Climate Engineering, http://www.spice.ac.uk/.
4. Paul Horn, "Notre Dame Receives Carnegie Classification for Community Engagement," press release, January 10, 2011, https://news.nd.edu/news/notre-dame-receives-carnegie-classification-for-community-engagement/.
5. EmNet is now part of Xylem. https://www.emnet.net/.
6. *Tiny Vampires* (podcast), http://www.tinyvampires.com/.
7. The Brain Scoop, YouTube channel, https://www.youtube.com/user/thebrainscoop.
8. Malawi Matters, http://www.malawimatters.org/.
9. University of Notre Dame, "Science Policy Initiative," https://sites.nd.edu/spi-club/.

References

Aristotle. 2014. *Nicomachean Ethics*. Translated by C. D. C. Reeve. Indianapolis: Hackett.
Baron, Marcia. 2010. *Escape from the Ivory Tower: A Guide to Making Your Science Matter*. Washington, DC: Island Press.
Bird, Stephanie J. 2014. "Social Responsibility and Research Ethics: Not Either/Or but Both." *AAAS Professional Ethics Report* 27, no. 2: 1–4.
Bourgeois, Mark. 2012. "Autonomy and Exploitation in Clinical Research: What the Proposed Surfaxin Trial Can Teach Us About Consent." *Ethics in Biology, Engineering & Medicine* 3:1–3, 51–56.

Bourgeois, Mark. 2017. "Design and Assessment of the Social Responsibilities of Researchers Graduate Training Program at the University of Notre Dame." Paper presented at the American Society for Engineering Education Annual Conference and Exposition, June.

Forge, John. 2008. *The Responsible Scientist: A Philosophical Inquiry*. Pittsburgh: University of Pittsburgh Press.

Hoskin, Keith. 1996. "The 'Awful Idea of Accountability': Inscribing People into the Measurement of Objects." In *Accountability: Power, Ethos and the Technologies of managing*, edited by R. Munro and J. Mouritsen, 265–82. London: International Thomson Business Press.

Jungk, Robert. 1958. *Brighter than a Thousand Suns: A Personal History of the Atomic Scientists*. Translated by James Cleugh. San Diego: Harcourt.

Kremer-Herman, Nathaniel. 2017. "Ethical Considerations of Doing Research at Scale." Poster presented at the Practice & Experience in Advanced Research Computing conference, July 9–13, 2017, New Orleans.

Lederach, Angela J. 2016. "Reflexivity, Responsibility and Reciprocity: Guiding Principles for Ethical Peace Research." *International Journal of Conflict Engagement and Resolution* 4, no. 1, 104–124.

MacIntyre, Alasdair. 2007. *After Virtue*. 3rd ed. Notre Dame, IN: University of Notre Dame Press.

Macrina, Francis L. 2005. *Scientific Integrity: Text and Cases in Responsible Conduct of Research*. 3rd ed. Washington, DC: American Society for Microbiology.

Muller, Jerry Z. 2018. *The Tyranny of Metrics*. Princeton: Princeton University Press.

National Academies Committee on Science, Engineering, and Public Policy. 2009. *On Being a Scientist: A Guide to Responsible Conduct of Research*. 3rd ed. Washington, DC: National Academies Press.

Pielke, Roger A., Jr. 2007. *The Honest Broker*. Cambridge: Cambridge University Press.

Russell, Daniel C. 2015. "Aristotle on Cultivating Virtue." In *Cultivating Virtue*, edited by Nancy Snow. Oxford: Oxford University Press, 17–48.

Shamoo, Adil E., and David B. Resnik. 2009. *Responsible Conduct of Research*. 2nd ed. New York: Oxford University Press.

Schwartz, Barry, and Kenneth Sharpe. 2010. *Practical Wisdom: The Right Way to Do the Right Thing*. New York: Riverhead Books.

van den Hoven, Jeroen, et al. 2013. "Options for Strengthening Responsible Research and Innovation." Expert Group on the State of Art in Europe on Responsible Research and Innovation, European Commission. EUR25766 EN.

14

Dynamic Boundaries

Using Boundary Work to Rethink Scientific Virtues

Louise Bezuidenhout and Dori Beeler

> "Science" is a cultural space: it has no essential or universal qualities.
> Rather, its characteristics are selectively and inconsistently attrib-
> uted as boundaries between "scientific" space and other spaces are
> rhetorically constructed.
>
> —Thomas Gieryn (1999, xii)

Modern life abounds with references to science. Newspapers trumpet discoveries
by scientists, or bemoan the detrimental effects of science on society. Television
shows offer images of gleaming cabinets, bubbling and highly colored liquids,
and "scientists" in white coats and safety glasses. Advertising campaigns make
extensive use of scientific-sounding vocabulary and often feature endorsements
by scientists. Through these different mediums we are given the idea of science
being a privileged practice conducted by "the scientist"—an individual somehow
special, privileged, and outside of normal society.

Regulators and policy developers are often little better. While recognizing the
heterogeneity of research practices and products, expediency often demands
that these multifarious divisions and subdivisions are collapsed into one over-
arching "science." We often see references to "science" and "the scientist" as
though they were somehow a coherent whole populated by similar practitioners
(e.g., National Academy of Sciences 2012). Thus, for the sake of efficiency, the variety (and messiness) of the landscape of scientific research is often overlooked.

While this undoubtedly makes discourse more efficient, it is important to
question what we are losing by underplaying the heterogeneity of scientific
knowledge production. What are the consequences of making science a "place-
less place" and the scientist a "faceless face"? In this chapter we work backward
from the idea of the scientist as somehow separate from society and governed by
a separate set of rules and expectations. Using key texts from the social studies
of science, we present a different interpretation of science—one that has no

Louise Bezuidenhout and Dori Beeler, *Dynamic Boundaries* In: *Science, Technology, and Virtues*. Edited
by: Emanuele Ratti and Thomas A. Stapleford, Oxford University Press. © Oxford University Press 2021.
DOI: 10.1093/oso/9780190081713.003.0015

absolute quality or essential characteristic but rather is defined by the ebb and flow of socially negotiated, dynamic boundaries. Using this representation of science, we question whether the scientist can be held apart, or whether they are simply a member of society engaged in a communal social activity. From a virtue ethics perspective, we then ask what this means for attempts to identify a specific set of virtues that are necessary and sufficient for scientific practice and the associated dynamic boundaries.

Being a Responsible Scientist

While many academics may laugh at the highly stylized representations of science and scientists in popular media, these categories nonetheless play an important role in the internal representations of science. While recognizing the vast heterogeneity of disciplines, practitioners, institutions, and national contexts, scientific communities regularly use the terms *science* and *scientists* as indicators of solidarity and unity (e.g., Australian Academy of Sciences et al. 2001; Fouchier et al. 2012). This is unsurprising, given the unifying force of the scientific method and its opposing push away from knowledge production in society/religion/traditional knowledge. For many scientists, there is indeed an ideal of science that is characterized by certain traits, or "essences."

Many sociologists, historians, and philosophers have attempted to outline the "essence" of science. This has led to extensive theories on the accuracy, utility, credibility, and authority of science (Merton 1942; Popper 1992; Kuhn 1962). Similarly, ethics has developed codes of conduct, schemas for responsible conduct of research (Macrina 2007), and descriptions of research misconduct (Martinson, Anderson, and de Vries 2005), all of which are premised on a central ideal of what science is. To do so, they reference science's seemingly unique, invariant, essential, and universal features. Recently, scholars working within the framework of virtue ethics (Pennock and O'Rourke 2017) have started to contribute to this project by attempting to identify key virtues necessary for the responsible practice of science.

These ethical representations of science and the scientist have been very influential in how we understand responsible and ethical research. They have served as a unifying force within scientific communities (via codes of conduct), commitments to the public in the face of scandals, and vocabularies of justification for funding, investment, and support. The prominent role means that their role in ethics is often uncritically accepted. Thus, while the heterogeneity of scientific research and practices is well known—and widely examined by the social studies of science—it is often underrepresented in ethics discussions in favor of more sweeping generalizations. This is problematic for a number of reasons. First, by being inherently decontextualized, they continue to struggle with the limitations of being content-poor (Engelhardt 1996) and must continually confront the ideal/real gap between value and implementation (Heimer 2013).

Perhaps more problematic, however, is that the continued use of these categories not only unifies disparate communities but others them from surrounding society. This "othering sets up a superior self/in-group in contrast to an inferior other/out-group, but this superiority/inferiority is nearly always left implicit" (Brons 2015, 70). *Othering* refers to the process of defining or labeling someone as being different in relation to another group or to a social norm. By using these categories, ethics discourse reinforces notions of the scientist as being set apart from society and of science as a practice set apart from other knowledge-producing activities.

An example of this othering is found in the use of the white lab coat. In the early twentieth century, medical doctors added the white coat to their attire in order to associate themselves with laboratory science; the goal was to "convey trust and to distinguish themselves from other hospital staff" (Salhi 2016, 208). This shift demonstrates how physicians, in identifying with the distinguishing features of scientists as other, use this feature to create a superior in-group within the hospital setting.

Understanding how othering is negotiated has become a prominent theme in the social sciences. Two strategies were put forward by anthropologist Claude Lévi-Strauss about how the other is dealt with by society: either boundaries are eliminated between the mainstream and the other, or the boundaries (including institutional boundaries) between them are fortified, which serve to exclude the other from society (Lévi-Strauss [1955] 2012). Consequently, the other is subjected not to mainstream social norms but to a different order of accountability, either privileged or disadvantaged.

In the case of science, the same phenomenon has occurred, with the scientist being set apart or othered. By trying to achieve too unified an image of what science is, science ethics has become trapped by images of science as a privileged occupation that occurs in specific locations. Similarly, the individuals occupying these spaces, scientists, have become othered as entities without personalities, backgrounds, or preferences. When one dons the white coat, othering takes place. All distinguishing features disappear in service of the "greater good" of knowledge creation. In this setting, the knowledge claims of science are presented as lacking connection to any social context (Oudshoorn 1996).

This othering has several consequences pertinent to science ethics. If the scientist is positioned as operating outside of the main body of society, it is difficult to provide consistent accounts of individual ethics, as it requires one individual to maintain two personas concurrently. Conversely, it does not provide insight into how and when one "becomes" a scientist. If someone *is* a scientist, is it only when that person dons a lab coat or steps into a lab, or is it anytime that person is thinking about science? Moreover, locating science within specific (and often highly stylized) spaces means that many ethics narratives fail to recognize that the spaces that constitute science are never static and have historically been subject to marked change.

Tracing the Outlines of Science

Since the mid-twentieth century, the social sciences have highlighted problems with the positivist vision of impartial science. A growing number of studies detail the social processes that act between nature and knowledge. Indeed, "interests, rhetorical tropes, power, identity, hands-on practices, tacit skills, instruments, experimental systems, and (as a catchall) culture are now standard ingredients in sociological studies of the construction of scientific knowledge" (Gieryn 1999, xii). This has opened up how science is analyzed, including the spaces in which science happens (Traweek 1988), the technology with which science is conducted (Rabinow 1996; Lynch 1985; Bijker, Hughes, and Pinch 2012), influences on scientists' worldview (Medin and Bang 2014), and the quality of character of the scientist (Deane-Drummond, Stapleford, and Narvaez 2018). Each of these works has moved deeper into understanding the ways in which science is a social practice.

In framing science as a social practice, these studies draw attention to *who* is involved in the *where*, *when*, and *how* science happens. Unlike positivist representations of science, where the setting acts as a passive backdrop to knowledge revelation, constructivist representations require detailed descriptions of research settings to understand how and why knowledge is created. Increasingly, historians, anthropologists, and sociologists are providing rich narratives of the plethora of ways in which scientific research has occurred.

The sociology of scientific knowledge has been highly influential in championing a social constructivist approach to understanding scientific practice and knowledge creation (Bloor 2004). These studies present an alternative view of science in which scientists (and their research environments) are actively engaged in knowledge production. In these studies "the social conditions and effects of science, and with the social structures and processes of scientific activity" are the object of study, and taken as the defining description of scientific practice (Ben-David and Sullivan 1975, 203).

Sociological investigations of science have been supported by a range of historical and philosophical studies, all of which have detailed the highly complicated social processes underpinning scientific research. In a recent book, Werrett (2019) studied the temporal evolution of research spaces, showing how they have evolved from domestic spaces using a combination of homemade and purchased equipment to the dedicated laboratories familiar in modern research. Similarly, other historical studies have detailed the temporal changes in both the funders and practitioners of science—from the gentleman scientists of the seventeenth century to the growing influence of commerce, the military, and civil funding (Hicks and Stapleford 2016).

Other historical studies have followed the production and cultural authority of technoscientific knowledge from the early part of the twentieth century to the late twentieth and early twenty-first centuries (Shapin 2008). Livingstone (2003) examined the geographies of science, showing how specific locations influence the making of scientific knowledge, and how local experience is transformed into shared generalizations. Modern anthropological studies, such as that by Traweek (1988), enrich understandings of cultural differences in working cultures and environments.

Yet more studies have outlined radical shifts in understandings of what constitutes research. For example, the development of genomics and epigenetics out of late twentieth-century genetics has involved radical conceptual changes in understandings of hereditary, health, and disease (Hine 2006). Similarly, the development of Big Data continues to cause radical changes in day-to-day research activities in many fields of biomedical and social science (Ankeny and Leonelli 2015; Leonelli 2014).

In recent years, the emergence of the citizen science movement (Heckler et al. 2018), together with the ever-evolving open-science landscape (Borgman 2015), has seen the practices of scientific knowledge production move beyond the traditional locations of research. By harnessing the resources available through open access, hardware, software, and data, research is being conducted by the public in makerspaces, community laboratories, public spaces, and private homes.

These research agendas all challenge the continued use of "science" and "the scientist" as useful categories for description. How can any single term do justice to the heterogeneity of the spaces, places, practices, and people involved in scientific knowledge production? Recognizing this heterogeneity does more than simply complicate attempts to use the term *science* as a catchall or as shorthand for a body of knowledge-producing practices; it also complicates attempts to generalize or privilege the practice of science. If science and scientists have been (and are) many different things, how are judgment and skills recognized, how is conflicting evidence assessed, and how is authority assigned and maintained?

Gieryn and the Boundaries of Science

Social studies of science not only detail the limitations of essentialism when demarcating the boundaries of science but also draw attention to the inherently dynamic nature of the divisions between science and society. Recognizing the ebb and flow of these boundaries necessarily calls into question the epistemic authority of science. Essential demarcation criteria (Merton 1942; Popper 1992; Kuhn 1962) can no longer explain the epistemic authority of science within these fluctuating settings.

Much of the research on the "social-ness" of science has focused on the creation of scientific knowledge—how social processes are intertwined with research practices. In the late 1990s, a new area of study emerged, looking not at this "upstream" account of knowledge production but rather at the "downstream" incorporation of knowledge into society. This field looked at how "scientists, their expertise, their claims and material artifacts eventually leave laboratories and technical journals and make their way out into the rest of the social world, where they are called upon to settle disputes, build airplanes, advise politicians, ascertain truth" (Gieryn 1999, ix)—in short, why science is trusted to provide credible and useful accounts of nature.

In 1999 Thomas Gieryn published a book entitled *Cultural Boundaries of Science: Credibility on the Line* which suggests that science is a "space on maps of culture, bounded off from other territories, labeled with landmarks showing travelers how and why it is different from regions of common sense, politics or mysticism. These cultural maps locate (that is, give a meaning to) white lab coats, laboratories, technical journals, norms of scientific practice, linear accelerators, statistical data, and expertise" (Gieryn 1999, x). The boundaries between science and other activities in this "cultural map" are not static, predetermined, or absolute, but rather are drawn and redrawn by dynamically engaged knowledge makers and users.

Gieryn suggests that the "cultural cartographies of science-in-culture are historical phenomena, with a local and episodic (rather than transcendent) existence. The same concatenation of interests, identities, discourses, and machineries that come together to make scientific knowledge have also come together to shape representations of science itself in a contextually contingent way" (Gieryn 1999, x). This representation of science foregrounds the dynamism of scientific borders, suggesting that the boundaries of science are continually formed and reformed depending on "who is struggling for credibility, what stakes are at risk, in front of which audiences, and what institutional arena" (Gieryn 1999, xi).

Gieryn's work liberates descriptions of science from any conception of essentialism. Instead, he suggests that what is "in" science depends on who draws the map. Thus, the main force controlling the demarcation between science and society is not essential epistemic authority but credibility. These boundaries are created by individuals inhabiting key social roles (Gieryn 1999, 22):

- Those who draw the boundaries by seeking to attach epistemic authority to their claims about nature

- Those who rely on these boundaries and knowledge maps as they allocate credibility in the course of making practical decisions based on the supposed reliable knowledge
- Those affected by the allocation of epistemic authority

In social interpretations of science, research has to win credibility just as any other social practice does. If science can no longer rely on being exempt from the need to win credibility, without such credibility there can be no product at all. Understanding social and cultural practices through which credibility is attributed is paramount (Shapin 1995, 257). If one accepts science as a social process, then the means through which it gains the epistemic authority that facilitates this downstream consumption is similarly social. Science becomes just one area in the communal epistemic landscape offering insight into the world around us.

It is easy to see how this approach contrasts directly with earlier positions of science that maintained that the "epistemic authority of science is *justified* (not just episodically won) by the unique, necessary and universal elements of its practice—behaviours, dispositions, methods, rules, tools and languages that simply work best to make truth" (Gieryn 1999, 25). The boundaries of science serve many important roles that underpin how we understand modern research, particularly in linking the evolution and maintenance (and defense) of boundaries with practical action (Mody 2011). Borders and territories of science facilitate the pursuit of goals and interests by cultural cartographers as well as by audiences and stakeholders (Gieryn 1999, 23). By conferring expertise and credibility, these boundaries play key roles in public support, funding, and institutional longevity.

Boundaries, Credibility, and Ethics

Gieryn's position offers a direct challenge to the descriptions of science discussed earlier in this chapter. By suggesting that there is no formal space "apart" in which science occurs, Gieryn suggests that scientists are just some of many knowledge producers on the cultural map. This directly counters the commonly held position that scientific knowledge creation is an activity apart from other forms of knowing and meaning-making. In turn, this necessitates that we question traditions of privileging scientific knowledge production and holding scientists to a different standard of accountability than other knowledge producers.

The tradition of setting scientists apart from other knowledge creators has significant implications for our interpretations of ethical conduct. These implications can be summarized in the following manner:

276 VIRTUES AND RESEARCH ETHICS

1. Scientific knowledge becomes privileged over other forms of knowledge. This creates biases against other forms of knowledge creation that also employ the scientific method, such as the social sciences.
2. The creation of scientific knowledge becomes lionized, and it is seen as a unique form of meaning-making requiring unique ethical standards.
3. The lionization of scientific knowledge and its prioritization over other forms of knowing causes a bias toward production over other forms of ethical behavior.

The position put forward by Gieryn and others erases the distinct divide between science and society. The activities of knowledge creation within scientific research are no longer ring-fenced off from other knowledge creation practices. Rather, they are part of a continuous cultural landscape in which communities engage in social negotiations to promote their knowledge contributions. It is important to note, however, that defining science by its boundaries instead of by any "essential qualities" need not detract from recognizing either the skills and expertise of scientists (Collins 2010) or the unique social structures of scientific communities (Traweek 1988). Indeed, understanding the power dynamics involved in boundary maintenance is undoubtedly assisted by framings of research as a communal practice with its associated internal and external goods. What is different, however, is that these internal and external goods are best understood as *contingent* on where the boundaries of science are drawn and by whom.

Positioning scientists not only as knowledge producers but also as asserters/ recipients of credibility complicates any attempts to other them into ivory towers. In the practices through which they negotiate or accept credibility they are no different from any other community of knowledge producers, such as citizen scientists, social scientists, or traditional knowledge practitioners. Their key roles and responsibilities therefore have to do with upholding the position of legitimacy that their field has in communal understanding, rather than being conduits of some kind of privileged knowledge revelation.

This does not necessarily mean that there are no values in science, however. In her book *Implicit Meanings*, Mary Douglas argues that the means through which a culture distributes credibility are intimately bound up with its moral life. Any account of a particular culture is in effect a description of the techniques through which credibility and risk are managed (Douglas 1999). In this way, the boundaries of science that demarcate credibility also become dynamic (re)negotiations of norms and values. As suggested by Mulkay (1976), this allows us to reframe norms and values from empirical descriptions of scientific practice to means of describing or justifying actions (Mulkay 1976).

Because boundaries may be understood as not only depicting but also sustaining cultural realities, they play an important ethical role in both science and society (Gieryn 1999, xii). They do not just represent a distinction between science and other communities (religion, engineering) but are also intimately intertwined with the moral fabric of society.

A Different Interpretation of Science as a Communal Practice

How, then, is it possible to depict the ethics of these fluctuating spaces of knowledge production? Virtue ethicists generally agree that science can be understood as a MacIntyrian practice, associated with internal and external goods. Hicks and Stapleford (2016) extend this notion of practice further, arguing for a communal practice in which communities of individuals are characterized by sustained activity; collaborative, goal-oriented, shared internal goods; and scale and temporal scope of activity.

The shared internal goods, or "goods of excellence" (Hicks and Stapleford 2016, 458), within a practice have intrinsic value; that is, they must have value in and for the practice itself, not for some other externally valued reason. For science these goods could include knowledge creation and the exercise of disciplinary skill. The goods of a practice are the goals that the practice sets out to achieve, and goods of excellence are achieved by engaging in a communal practice in such a way as to benefit the practice itself (MacIntrye 2007). External goods, on the other hand, are finite and are obtained and valued only by and for the individual (for example, money, prestige, or fame).

The external goods of science are thus highly dependent on where the boundaries of science are drawn. While many virtue ethicists recognize that "the boundaries of communal practices are established sociologically and politically" (Hicks and Stapleford 2016, 456), the boundary politics of Gieryn and others are often overlooked. As a result, virtue ethicists suggest that future outcomes resulting from interactions of accountability will be instrumental in defining the boundaries of the practice. Here we suggest that the use of static boundaries between science and society is particularly problematic. Overlooking the highly social way in which scientific boundaries are continually formed and re-formed results in a tendency to overlook the key role that every scientist has in continually mediating and maintaining the practice as a whole. This can lead virtue ethics to overlook the politics informing and shaping relationships individuals have within the group and outside the group as well as relationships between one practice and another.

Porous Boundaries and Scientific Virtues

In the previous sections we have presented a case for science as a part of the cultural cartography of knowledge production. What distinguishes science from other forms of knowledge production, we argue, is not a set of essential criteria but rather the dynamic social practices whereby boundaries are negotiated and renegotiated with society. So, we must ask, what does that mean for our understanding of virtues of this reimagined communal practice?

The absence of a set of essential criteria describing science suggests that attempting to identify "essential virtues" associated with scientific knowledge production may be a moot point. Rather, the virtues underpinning responsible scientific knowledge production are the same as (or very similar to) those seen in many other activities within the knowledge production landscape. Compiling lists of "essential virtues" runs the risk of overlooking important synergies between science and other forms of knowledge production—particularly in the modern milieu, where the boundaries between science, industry, and society are increasingly blurred.

This is not to say that these lists cannot be useful. Because the virtues are involved in robust knowledge production, identifying them can be an important teaching tool and guideline. However, the temptation to view them as descriptions of the distinguishing characteristics of science, in contrast to other forms of knowledge production (as many purport to do), means that such lists of virtues can be alienating rather than unifying. Allowing them to become a vocabulary of justification for scientific behavior also masks some key responsibilities of scientists to their colleagues and society (Bezuidenhout 2017).

Recognizing the virtues involved in responsible knowledge production is important for understanding how individuals achieve excellence in science as a practice. Nonetheless, in MacIntyre's account of practice, virtues also serve to protect the practice from the threat of corruption by goods of efficiency, and they serve as constitutive components of the good human life (MacIntyre 2007). Thus, virtues consolidate and stabilize science as a practice. These functions, to return to the discussion on credibility, are the same ones that establish, maintain, and defend the dynamic boundaries that demarcate science on the cultural landscape.

By recognizing the important role that boundaries play not only in demarcating science from other knowledge-producing activities but also in defining the practice of science, we can acknowledge the possibility that the virtues that really distinguish science are those involved in boundary work. The question thus shifts from "What virtues make the practice of science different from other knowledge-producing activities?" to "What virtues are needed to responsibly mediate the boundaries between science and society?"

Responsibly Establishing, Maintaining, and
Defending Boundaries

"The stakes—authority, jobs, fame, influence, nature—create big incentives for some cultural cartographers to (re)draw the boundaries of science one way, just as others have good reason to counter with maps of their own" (Gieryn 1999, 15). This is undoubtedly true for science, where the stakes of being a legitimized knowledge producer are high. As national research budgets are put under strain, competition for existing grants increases, and the possibility of tenured jobs decreases (Arnold 2014; Anderson et al. 2007; Herbert et al. 2014), the stakes of drawing boundaries have never been higher.

Being in a position to influence the boundaries of science has significant implications not only for how science is positioned within the cultural landscape but also for how it is internally organized as a practice. These boundaries control the entry and exit of individuals, determine the conditions of work, and dictate the socialization processes with which new initiates are integrated into the practice. All of these activities confer significant status on the individuals who manage the boundaries, making them the trendsetters and the ones to follow.

Recognizing the power of boundaries makes it easy to see how boundary work is an activity with significant ethical implications. Boundary making, maintenance, and defense can easily be opened to abuse as individuals seek to promote personal gain over the goods of excellence or communal and societal goods. The activities involved in boundary work are multiple and varied, and trying to limit them to a list is probably unhelpful. Nonetheless, we would view these activities as hinging on scientists being aware of their own discipline, the needs of the community, and their position in society. They would include engagement with the public—not only through formal channels but also in the micro interactions of daily life.

A virtuous scientist working in the fluctuating world of science would necessarily be involved in ensuring the honest policing and maintenance of boundaries through the expulsion of certain types of scientific influences deemed problematic, the expansion of science's areas of influence in the larger world, and protection of the authority of science as a practice. It is probable that many of the boundary-maintenance activities would require honor, tenacity, perseverance, and courage to defend the use of particular types of knowledge over other forms of knowing. Nonetheless, successful boundary discourse also includes tolerance for other stakeholder opinions, and the acceptance that others' preferences and ideas may differ from those held by scientists yet are not necessarily wrong or invalid. Similarly, a degree of humility is required for the scientist to recognize that science is not the only form of knowledge production, but is privileged to be accorded such high prestige.

Many boundary negotiations happen in situations with high stakes or diametrically opposed positions. In such situations, it is easy for discussion to become emotionally charged, as discussants feel themselves to be under attack, undervalued, or unfairly criticized. This can lead to attempts to discredit or muzzle opposing positions—sometimes using rather underhanded techniques. Recognizing the possibility of such difficulties highlights how a responsible and virtuous scientist must be able to rise above such temptations. Exercising temperance in boundary negotiations so as not to abuse the practice's privileged position is an extremely important element of responsible science. Similarly, having the patience to remain calm and not become annoyed when dealing with these difficult debates is vital. Such patience also extends to committing to long-term engagement without becoming bored or losing interest. Virtuous scientists recognize the difference between winning a specific debate and undermining public trust in science by damaging lines of communication.

The effective negotiation of scientific boundaries means that science—and scientists—will receive credibility from the public. It is important to note that giving/acquiring credibility is not a right linked to any essential criteria of science but rather is a privilege. As mentioned by Gieryn, "there is nothing on a scientific instrument, a fact, a statistic, or a white lab coat that says 'true' or 'trustworthy' or 'credible'" (Gieryn 1999, 19). Thus, the epistemic authority of science is enacted "as people debate (and ultimately decide) where to locate the legitimate jurisdiction over natural facts" (Gieryn 1999, xiii).

Being in receipt of public trust comes with significant responsibility. By being afforded credibility for the knowledge they produce, scientists are given a considerable amount of cultural authority. They are also in receipt of public resources that enables them to dictate societal change. The activities facilitated by credibility (such as the production of truth, the giving of advice, etc.) therefore have a temporal component, as they have a direct impact on cultural futures. It is therefore of considerable importance that the scientists who benefit from being afforded this credibility exercise foresight in all their activities. It is only through being actively engaged in considering the consequences of current actions that scientists will truly be worthy of the credibility they are given.

Developing "Boundary Virtues"

Understanding the role of boundaries affords a very different interpretation of what science is. Instead of a static space with certain essential practices/traits, science is defined by what its ever-fluctuating boundaries enclose or exclude. This presents an interesting challenge to current ethics discourse, as there is no option to predetermine the context and codify the rules. Ethics must become flexible and contextual in order to cope with this challenge. In this, as discussed earlier, virtue ethics can make a significant contribution.

In response to these challenges, we advocate for a new understanding of the scientist as a morally robust individual who is able to flexibly deal with the fluctuating spaces created by boundaries without losing ethical integrity. This approach links with many contemporary virtue ethicists, who are arguing that "the moral agent in her role as scientist should not be thought of as separate from the moral agent as a person. Indeed, a key component of virtue ethics is that the moral agent has integrity within her character, that her emotions, motivations, and reasons for a particular action align together" (Gottlieb 1994, 288). Therefore, "the moral agent's various social roles and how she carries them out should cohere and are broadly illustrative of who she is as a person" (Chen 2015, 79).

Defining science by its boundaries rather than by any "essential criteria" has significant implications for science ethics. In particular, it opens up the possibility of revisiting how scientists should act on their communal responsibilities to their peers, students, and employers. Current ethics discourse is hampered by the tendency to focus on misconduct rather than on being a good community member. Framing science as one of many communal practices allows a stronger focus on how socially held norms transition into science spaces and what should—or should not—be condoned. Should Watson and Crick be lionized as exemplary scientists, or should they be criticized for the uncredited use of Franklin's work? How would such behavior be judged in normal society if Watson and Crick were business owners or members of a social club? Would their goods of excellence within a communal practice have been viewed differently if they had allowed Franklin to be recognized for her work?

Understanding science by its ever-changing boundaries also enables scientific ethics to confront some areas of modern knowledge production that have previously been problematic. The rise of the citizen science movement in the last ten years has seen the practice of science extend out of traditional research settings and into community laboratories and non-academic spaces (Heckler et al. 2018). Moreover, the scientists working in this movement have not necessarily received formal science training. Nor do they necessarily publish or communicate in traditional academic avenues, such as journals, conferences, and formal collaborations.

Citizen science thus presents a tricky problem for the more traditional approaches to science ethics. None of the previously assumed characteristics of responsible science, such as ethics training and oversight, institutional support, and community scrutiny, can be said to necessarily hold in these spaces. Moreover, because these laboratories/communities evolve specifically *not* to be academic research spaces, attempting to transpose academic ethics oversight to these spaces wholesale is inappropriate and likely to be destructive.

Despite the obvious (and intentional) differences between the citizen science communities and more traditional science spaces, there is much overlap. Indeed, collaborations between these two knowledge-producing communities are getting increasingly common, while large crowdsourcing platforms such as Zooniverse are drawing the general public into research practices. In these areas we see a contemporary renegotiation of the boundaries of science. Understanding how these boundaries are redefined and how communities form and reform within this changing landscape will be of considerable benefit for ethicists starting to work in these areas.

Similarly, in recent years there has been a considerable rise in the reliance on "evidence-based policy" and widespread data collection to inform decision-making. In this way, practices of experimenting, testing, and measuring extend far beyond the boundaries of the laboratory or university. "Ubiquitous testing has entered our everyday lives and homes: our bodies are regularly medically tested; air, water, and food quality are continually tested for pollutants; toys are tested for chemicals; schools and teacher performance are tested; and even banks are tested in scenario simulations" (Beisel, Calkins, and Rottenburg 2018, 109). By redrawing the boundaries of who collects evidence and how this evidence is used, we are actively redrawing the boundaries of credibility and authority.

Concluding Comments

Defining science by its boundaries, rather than by any essential criteria, marks a move away from traditional scientific ethics. Instead of trying to find ways in which to separate out all scientific knowledge production from other forms of meaning-making, the concept of "boundary ethics" refocuses attention on why and when scientists and society draw these distinctions. Science, it would seem, is never going to be something that can be consistently described by a single set of criteria; similarly, our ethics of science needs to be flexible enough to deal with this temporal, geographical, and social variation.

In doing so, boundary ethics offers a new way of attempting to define what constitutes "responsible science practice." Instead of offering a separate moral code to define scientists' activities in spaces outside of science, it focuses attention on the responsibilities of negotiating and maintaining credibility. The honor of being afforded credibility by society is often underplayed in science ethics, and boundary ethics repositions this trust as the central component of responsible science practice.

While we recognize that this short chapter cannot fully outline either boundary ethics or provide a road map for how to study it, we hope that by drawing attention to these issues this chapter will stimulate interest in a virtue ethics that draws more closely on social studies of science. We believe that it is only through understanding science as an infinitely variable social practice that virtue ethics can realize its potential contributions to science ethics.

References

Anderson, Melissa S., Emily A. Ronning, Raymond De Vries, and Brian C. Martinson. 2007. "The Perverse Effects of Competition on Scientists' Work and Relationships." *Science and Engineering Ethics* 13, no. 4: 437–61.

Ankeny, R. A., and S. Leonelli. 2015. "Valuing Data in Postgenomic Biology: How Data Donation and Curation Practices Challenge the Scientific Publication System." In *Postgenomics*, edited by S. S. Richardson and H. Stevens, 126–49. Durham, NC: Duke University Press.

Arnold, Carrie. 2014. "The Stressed-Out Postdoc." *Science*, July.

Australian Academy of Sciences, Royal Flemish Academy of Belgium for Sciences and the Arts, Brazilian Academy of Sciences, Royal Society of Canada, Caribbean Academy of Sciences, Chinese Academy of Sciences, French Academy of Sciences, et al. 2001. "The Science of Climate Change." *Science* 292, no. 5520: 1261.

Beisel, Uli, Sandra Calkins, and Richard Rottenburg. 2018. "Divining, Testing, and the Problem of Accountability." *HAU: Journal of Ethnographic Theory* 8, nos. 1–2: 109–13.

Ben-David, J., and Sullivan, T. A. 1975. "Sociology of science." *Annual Review of Sociology*, 1(1), 203–222.

Bezuidenhout, L. 2017. "The Relational Responsibilities of Scientists: (Re)Considering Science as a Practice." *Research Ethics* 13, no. 2: 65–83.

Bijker, Wiebe E., Thomas Parke Hughes, and T. J. Pinch. 2012. *The Social Construction of Technological Systems: New Directions in the Sociology and History of Technology.* Cambridge, MA: MIT Press.

Bloor, David. 2004. "Sociology of Scientific Knowledge." In *Handbook of Epistemology*, edited by I. Niiniluoto, M. Sintonen, and J. Woleński, 919–962. Dordrecht: Springer.

Borgman, Christine L. 2015. *Big Data, Little Data, No Data: Scholarship in the Networked World.* MIT Press.

Brons, Lajos L. 2015. "Othering: An Analysis." *Transcience: A Journal of Global Studies* 6, no. 1: 69–90.

Chen, Jiin-Yu. 2015. "Virtue and the Scientist: Using Virtue Ethics to Examine Science's Ethical and Moral Challenges." *Science and Engineering Ethics* 21, no. 1: 75–94.

Collins, H. M. 2010. *Tacit and Explicit Knowledge.* Chicago: University of Chicago Press.

Deane-Drummond, Celia, Thomas A. Stapleford, and Darcia Narvaez. 2018. *Virtue and the Practice of Science: Multidisciplinary Perspectives.* Notre Dame, IN: Center for Theology, Science, and Human Flourishing, University of Notre Dame.

Douglas, Mary. 1999. *Implicit Meanings: Selected Essays in Anthropology.* London: Routledge.

Engelhardt, H. Tristram. 1996. *The Foundations of Bioethics.* New York: Oxford University Press.

Fouchier, R. A. M., A. Garcia-Sastre, Y. Kawaoka, W. S. Barclay, N. M. Bouvier, I. H. Brown, I. Capua, et al. 2012. "Pause on Avian Flu Transmission Research." *Science* 335, no. 6067: 400–401.

Gieryn, Thomas F. 1999. *Cultural Boundaries of Science: Credibility on the Line.* Chicago: University of Chicago Press.

Gottlieb, Paula. 1994. "Aristotle on Dividing the Soul and Uniting the Virtues." *Phronesis* 39, no. 3: 275–90.

Heckler, Susanne, Muki Haklay, Anne Bowser, Zen Makuch, Johannes Vogel, and Aletta Bonn, eds. 2018. *Citizen Science: Innovation in Open Science, Society and Policy.* London: UCL Press.

Heimer, Carol A. 2013. "'Wicked' Ethics: Compliance Work and the Practice of Ethics in HIV Research." *Social Science and Medicine* 98: 371–78.

Herbert, Danielle L., John Coveney, Philip Clarke, Nicholas Graves, and Adrian G Barnett. 2014. "The Impact of Funding Deadlines on Personal Workloads, Stress and Family Relationships: A Qualitative Study of Australian Researchers." *BMJ Open* 4, no. 3: e004462.

Hicks, Daniel J., and Thomas A. Stapleford. 2016. "The Virtues of Scientific Practice: MacIntyre, Virtue Ethics, and the Historiography of Science." *Isis* 107, no. 3: 449–72.

Hine, Christine. 2006. "Databases as Scientific Instruments and Their Role in the Ordering of Scientific Work." *Social Studies of Science* 36, no. 2: 269–98.

Kuhn, Thomas. 1962. *The Structure of Scientific Revolutions.* Chicago: University of Chicago Press.

Leonelli, Sabina. 2014. "What Difference Does Quantity Make? On the Epistemology of Big Data in Biology." *Big Data & Society* 1, no. 1: 1–11.

Lévi-Strauss, Claude. (1955) 2012. *Tristes Tropiques.* Translated by John and Doreen Weightman. London: Penguin Books.

Livingstone, David N. 2003. *Putting Science in Its Place: Geographies of Scientific Knowledge.* Chicago: University of Chicago Press.

Lynch, Michael. 1985. *Art and Artefact in Laboratory Science: A Study of Shop Work and Shop Talk in a Laboratory.* London: Routledge & Kegan Paul.

MacIntyre, Alasdair C. 2007. *After Virtue: A Study in Moral Theory.* Notre Dame, IN: University of Notre Dame Press.

Macrina, Francis L. 2007. "Responsible Conduct of Research: Codes, Policies, and Education." *Academic Medicine* 82, no. 9: 865–69.

Martinson, Brian C., Melissa S. Anderson, and Raymond de Vries. 2005. "Scientists Behaving Badly." *Nature* 435, no. 7043: 737–38.

Medin, Douglas, and Megan Bang. 2014. *Who's Asking? Native Science, Western Science, and Science Education.* Cambridge, MA: MIT Press.

Merton, R. K. 1942. *The Normative Structure of Science.* Chicago: University of Chicago Press.

Mody, Cyrus. 2011. *Instrumental Community: Probe Microscopy and the Path to Nanotechnology.* Cambridge, MA: MIT Press.

Mulkay, Michael. 1976. "Norms and Ideology in Science." *Social Science Information* 15: 637–56.

National Academy of Sciences. 2012. *On Being a Scientist: A Guide to Responsible Conduct in Research.* 3rd ed. Washington, DC: National Academies Press.

Oudshoorn, Nelly. 1996. "A Natural Order of Things? Reproductive Science and the Othering of Things." In *FutureNatural: Nature, Science, Culture*, edited by G. Robertson, M. Mash, J. Bird, B. Curtis, L. Tickner, and T. Putnam, 122–32. London: Psychology Press.

Pennock, Robert T., and Michael O'Rourke. 2017. "Developing a Scientific Virtue-Based Approach to Science Ethics Training." *Science and Engineering Ethics* 23, no. 1: 243–62.

Popper, Karl R. 1992. *The Open Society and Its Enemies*. Princeton: Princeton University Press.

Rabinow, Paul. 1996. *Making PCR: A Story of Biotechnology*. Chicago: University of Chicago Press.

Resnik, David. 2005. *The Ethics of Science: An Introduction*. London: Routledge.

Salhi, Bisan. 2016. "Beyond the Doctor's White Coat: Science, Ritual, and Healing in American Biomedicine." In *Understanding and Applying Medical Anthropology: Biosocial and Cultural Approaches*, 3rd ed., edited by Peter J. Brown and Svea Closser, 204–12. London: Routledge.

Shapin, Steven. 1995. "Cordelia's Love: Credibility and the Social Studies of Science." *Perspectives on Science* 3, no. 3: 255–75.

Shapin, Steven. 2008. *The Scientific Life: A Moral History of a Late Modern Vocation*. Chicago: University of Chicago Press.

Traweek, Sharon. 1988. *Beamtimes and Lifetimes: The World of High Energy Physicists*. Cambridge, MA: Harvard University Press.

Werrett, Simon. 2019. *Thrifty Science: Making the Most of Materials in the History of Experiment*. Chicago: University of Chicago Press.

Index

For the benefit of digital users, indexed terms that span two pages (e.g., 52–53) may, on occasion, appear on only one of those pages.

Figures are indicated by *f* following the page number; 'n.' after a page number indicates the endnote number.